Encyclopedia of Medicinal Plants

ENCYCLOPEDIA OF MEDICINAL PLANTS

Edited by:

Esha Rami

Delve
PUBLISHING

www.delvepublishing.com

Encyclopedia of Medicinal Plants

Esha Rami

Delve Publishing
224 Shoreacres Road
Burlington, ON L7L 2H2
Canada
www.delvepublishing.com
Email: orders@arclereducation.com

ISBN: 978-1-77469-379-7 (Hardcover)

ABOUT THE EDITOR

Dr. Esha Rami is presently working as an Assistant Professor, in the Department of Life science, Parul Institute of Applied Science, Parul University, India. She Did her Post graduated from Ganpat University, Ph.D. in biotechnology from Hemchandracharya North Gujarat University, in the year 2015. She has authored a number of national and international publications in reputed journals.

TABLE OF CONTENTS

List of Figures .... *ix*
List of Abbreviations .... *xv*
Abstract .... *xvii*
Preface .... *xix*

A **1**

B **27**

C **38**

D **52**

E **66**

F **78**

G **87**

H **102**

I **112**

J **121**

K **133**

L **149**

M **171**

N **185**

O **191**

P........203

Q........222

R........231

S........252

T........269

U........281

V........287

W........297

X........316

Y........321

Z........335

Bibliography........345

Index........351

LIST OF FIGURES

Figure 1: *Abrus Precatorius*

Figure 2: Rosary Pea Dosage

Figure 3: *Acacia Arabica*

Figure 4: *Adhatoda Vasica* (Vaska)

Figure 5: The Onion

Figure 6: Garlic

Figure 7: Stone Age Andrographis Paniculata Leaves

Figure 8: Custard Apple

Figure 9: *Yucca Filamentosa*

Figure 10: *Capparis Tomentosa*

Figure 11: File:*Bassia latifolia*(Latin),Mahua Tree.JPG

Figure 12: Horticultural Impex Natural Bixa Orellana Seeds

Figure 13: *Brassica Juncea*

Figure 14: Catechu

Figure 15: Daikon juice, as well as the roots and leaves of the daikon plant, contain anti-inflammatory chemicals that aid to reduce inflammation throughout the body. This has a lot of beneficial impacts for your body, including maintaining heart health, lowering the risk of developing arthritis, treating gout, and reducing the discomfort and suffering associated with accidents and muscle strains.

Figure 16: The leaves are used to aid digestion and boost appetite. The blossom of the dandelion is high in antioxidants. Dandelion may also aid in the improvement of one's immune system. Dandelion root and leaves are used by herbalists to detoxify the liver and gallbladder, while dandelion leaves are used to support kidney function.

Figure 17: Dong quai advantages have long been recognized throughout Asia, where the root is also known as *Radix angelicae sinensis* and has been used as an everyday tonic and herbal medicine for generations.

Figure 18: *Desmodium triflorum* is a well-known and widely used medicinal plant in Sri Lanka. Its antioxidant, anti-inflammatory, anti-convulsant, and analgesic properties have been proven in recent pharmacological research. Alkaloids abound in this plant.

Figure 19: Cowpeas are high in the mineral (Iron), which helps to eliminate anemia. Iron aids in protein metabolism, which is necessary for the formation of RBCs and hemoglobin, as well as preventing anemia. Anemia is caused by a deficiency in hemoglobin and red blood cells.

Figure 20: Licorice root is now widely used to treat a variety of diseases, including heartburn, acid reflux, coughs, hot flashes, and viral and bacterial and infections. It's commonly sold as a pill or a liquid supplement.

Figure 21: Finger millet not only aids in maintaining good health and warding off ailments, but it also works wonders in preserving skin's youth and vitality by preventing premature aging. It can be used as the most powerful natural anti-aging beverage.

Figure 22: For various administrative routes, solutions can be developed. Orally: Drops, elixirs, and syrups, IV, IM, SC (parenterally) Mouthwashes, gargles, and throat sprays are used in the mouth and throat. Douches, ear drops, enemas, and nasal sprays are used in the body cavities.

Figure 23: Cloves are a great treatment for stomach disorders. These are beneficial in the treatment of stomach gas, nausea, diarrhea, vomiting, and other digestive system issues. Cloves are also expectorant and effective in cough treatment.

Figure 24: Essential oils are most known for their health advantages, which include calming allergy symptoms, controlling obstructed menstrual cycles, decreasing body odor, avoiding hair loss, digestion, recovering from colds and asthma, and assisting those with breathing problems.

Figure 25*: Oenothera Biennis* is the scientific name for the evening primrose plant. They are members of the Onagraceae family. Other names for it include German rampion, hogweed, fever-plant, and Kind's cure-all. The plant is edible and has been used for a variety of health advantages over time. Salads are garnished with the delicious blooms, while the immature seedpods are cooked. The leaves are then boiled. The plant is mostly found in North America's central and eastern regions.

Figure 26: Weak digestion, antispasmodic, intestinal parasites, bronchitis, flatulence, stomachache, epilepsy, ulcer, asthma, whooping cough, and influenza are among the disorders for which asafoetida is traditionally used. Asafoetida is a helpful treatment for a variety of stomach ailments.

Figure 27: According to studies, fumitory can aid with gallbladder disease, liver inflammation, and bile flow from the gallbladder. It has been demonstrated to increase liver function by up to 80%, as well as detox and heal the liver in general... as well as boosting liver activity. Fumitory also works as an antispasmodic for the gallbladder and reduces the production of gallbladder stones.

Figure 28: Garlic can be grown in both large scale and small scale farms

Figure 29: There are a variety of garlic oriented home remedies used in treating a number of diseases

Figure 30: Garlic contains heart-protecting elements

Figure 31: Garlic is processed to produce tablets and capsules

Figure 32: *Gardenia gummifera* plants produce resins used in treating different diseases

Figure 33: Properties of Gossypium sp contribute to its medicinal value

Figure 34: *Gmelina arborea* is native to the Indian sub-continent
Figure 35: GERD is linked to a condition called hiatal hernia
Figure 36: *Helianthus annus* is among the different types of helianthus
Figure 37: *Hibiscus rosa-sinesis* is native to China
Figure 38: Seeds of *Hordeum sativum*
Figure 39: HRT controls production of hormones
Figure 40: Juniper Berries.
Figure 41: Jimbu.
Figure 42: Cinder Jalapeno Pepper.
Figure 43: Kale.
Figure 44: Kava Kelp.
Figure 45: Lavender can be distilled into an essential oil, and has a range of medical applications.
Figure 46: Lemon.
Figure 47: Lemon balm extract.
Figure 48: Lemon Myrtle Extract.
Figure 49: Licourice root.
Figure 50: Marvel grass
Figure 51: Mahua plant (*Madhuca Longifolia*)
Figure 52: *Mentha arvensis*
Figure 53: the sensitive plant
Figure 54: flowering of magnolia champaca
Figure 55: Mustard plant
Figure 56: *Murraya Koenigii*
Figure 57: *Myristica fragrans*
Figure 58: *Nicotiana tabacum*
Figure 59: Neem Tree
Figure 60: the stinging nettle
Figure 61: The *Ochna Serrulata*
Figure 62: The Holy Basil
Figure 63: *Oryza Sativa*
Figure 64: Olive Leaves
Figure 65: The Common Bulb Onion
Figure 66: Ooregano

Figure 67: Rice bean (*Vigna umbellata*) before and after harvesting

Figure 68: Black pepper fruits are the source of one of the world's most widely and frequently used spices. Black, white and green peppercorns all come from Piper nigrum and are popularly used as a hot and pungent spice for flavouring food. Black pepper is also used in traditional medicine, particularly for digestive ailments.

Figure 69: Guava flower in Kabwe, Zambia

Figure 70: N-methyl-D-aspartate (NMDA) receptor activation by glutamate leads to phospholipase A2 (PLA2) activation and the generation of platelet-activating factor (PAF) and of arachidonic acid. PAF is synthesized through other metabolic routes as well [119], and elicits its actions through a PAF receptor. PAF activates COX-2 gene expression through protein kinase cascades and transcription factors. The COX-2 promoter is also a target for cytokines (activation) and glucocorticoids (inhibition). COX-2 protein then catalyzes the conversion of arachidonic acid into prostaglandin H2 (PGH2), the precursor of eicosanoids. The constitutive activity of COX-1 also catalyzes this metabolic step.

Figure 71: The prostaglandins are made up of unsaturated fatty acids that contain a cyclopentane (5-carbon) ring and are derived from the 20-carbon, straight-chain, polyunsaturated fatty acid precursor arachidonic acid.

Figure 72: Signaling through the G12 subfamily (comprising Gα12 and Gα13) transduces GPCR signals into Rho activation, actin remodeling, assembly of focal adhesions, and tyrosine phosphorylation of the focal adhesion-associated proteins FAK, CAS, and paxillin and complex formation between FAK and Src. Notice that FAK integrates multiple signal transduction pathways by multi-site tyrosine (e.g., Tyr397, Tyr576/577, Tyr861, and Tyr925) and serine (Ser722, Ser843, Ser910) phosphorylation. These pathways are implicated in both cell proliferation and cell migration. See text for details.

Figure 73: *Quebracho colorado* (*Schinopsis balansae*) wood

Figure 74: Queen's Delight

Figure 75: Leaves and flowers (male flowers on top)

Figure 76: Raspberry Ketone

Figure 77: Rhodiola has long been known as an adaptogen, a natural substance that increases your body's resistance to stress in non-specific ways.

Figure 78: Rose hips from *Rosa rugosa* (beach rose)

Figure 79: Developing queen larvae surrounded by royal jelly

Figure 80: Sandalwood essential oil works wonders as an antiseptic. It is suitable for both internal and external use. When taken orally, it aids in the prevention of infection in internal wounds and ulcers.

Figure 81: Ashoka, also known as *Saraca indica*, is a rain forest tree native to India's Deccan Plateau and Western Ghats. It blooms in clusters of orange and yellow flowers that turn crimson before falling.

Figure 82: Comfrey has been found to contain pyrrolizidine alkaloids (PAs), which are chemicals that have been linked to liver damage in humans. Over 600 distinct PAs have been discovered in over 6,000 different plants.

Figure 83: Foster is the president of Steven Foster Group, Inc., a company that offers consultancy, editorial, and stock photography services.

Figure 84: Sesquiterpene lactones, a diverse group of terpenoids identified from Asteraceae species, have a wide range of biological effects, and several of them, such as artemisinin, are now commercially available.

Figure 85: SSRIs are antidepressants that work by boosting serotonin levels in the brain. Serotonin is a chemical messenger (neurotransmitter) that transmits impulses between nerve cells in the brain (neurons). SSRIs prevent serotonin from being reabsorbed (taken back up) into neurons. This increases the amount of serotonin accessible, which improves message transmission between neurons. The selective nature of SSRIs comes from the fact that they only impact serotonin and not other neurotransmitters.

Figure 86: Tamarind leaf extracts have anti-oxidant properties in the liver and are used in cardiac and blood sugar-lowering medications. Young leaves can be used in rheumatism fomentation, administered to sores and wounds, or applied as a poultice to reduce swelling and relieve discomfort in joints. The leaves are used to make a sweetened decoction that can help with throat infections, coughs, fevers, and even intestinal worms.

Figure 87: *Terminalia chebula* (Combretaceae) is commonly used in Indian and Iranian traditional medicine to cure ailments such as dementia, constipation, and diabetes. The fruit of this tree, known as halileh or halilaj in Iranian traditional medicine (ITM), is utilized to produce cures.

Figure 88: A health service that has been used to prevent, detect, and treat illness for thousands of years. It is founded on the notion that qi (the body's vital energy) travels through the body's meridians (channels) and maintains a person's emotional, spiritual, intellectual, and physical well-being. Traditional Chinese medicine tries to rebalance and harmonize the body's natural opposing energies of yin and yang, which can impede qi and lead to disease.

Figure 89: Tumor necrosis factor is not normally detectable in healthy people, but it is detected at high amounts in the blood and tissues in inflammatory and infectious disorders, and serum levels are linked to infection severity.

Figure 90: It's used to manage menstrual irregularities, menstrual cramping, and leucorrhoea, as well as a range of other conditions affecting the male and female reproductive organs. It can help with impotence, genito-urinary weakness or irritation, as well as liver and kidney disorders.

Figure 91: Usnea is a lichen that can be found on trees. Lichens may appear to be solitary plants, but they are actually a mixture of fungus and algae that coexist for mutual benefit. Lichens grow in flat, bright patches.

Figure 92: Valerian Plant

Figure 93: Vitex Angus Castus plant

Figure 94: *Vernonia Anthelmintica* plant

Figure 95: Cowpea, *Vigna sinensis*

Figure 96: Vitex negundo
Figure 97: walnut tree
Figure 98: The wasabi plant
Figure 99: Watercress plant
Figure 100: Wheatgrass plant and juice
Figure 101: White oak tree
Figure 102: Wild Yam
Figure 103: Wormwood plant
Figure 104: *Withania somnifera*
Figure 105: *Xylia xylocarpa*
Figure 106: *Xylopia aethiopica*
Figure 107: *X. heucherella*
Figure 108: Both the leaves and flowers of Yarrow have medicinal value
Figure 109: *Yerba mate* is used to make herbal tea
Figure 110: *Ylang ylang* oil is a common essential oil for Ylang ylang
Figure 111: Mifoil Yarrow is of the Asteraceae family.
Figure 112: The roots of a Yellow dock plant are used for their medicinal value
Figure 113: Yellow Jessamine plants are known for their yellow flowers
Figure 114: The fruits of Yerba mansa are spiky in nature
Figure 115: Yohimbine is an active element in most Yohimbe products
Figure 116: The roots of Yucca have strong medicinal value
Figure 117: Maize is a good source of carbohydrates
Figure 118: *Zingiber officinale* is known to treat nausea and pain
Figure 119: Zinc is a crucial element used in the body's development
Figure 120: Zedoary is commonly grown in Thailand

LIST OF ABBREVIATIONS

AMP	Adenosine monophosphate
ALA	Alpha-linolenic acid
AHH	Aryl hydrocarbon hydroxylase
DGL	Deglycyrrhizinated licorice
ED	Erectile dysfunction
EPO	Evening primrose oil
FDA	Food and Drug Administration
GABA	Gamma-amino-butyric acid
GLA	Gamma-linolenic acid
GRAS	Generally recognized as safe
HSV	Herpes simplex virus
HRT	Hormone replacement therapy
IBD	Inflammatory bowel disease
MUFAs	Monounsaturated fatty acids
NSAIDs	Nonsteroidal anti-inflammatory drugs
PMDD	Premenstrual dysphoric disorder
PMS	Premenstrual syndrome
PGs	Prostaglandins
PAs	Pyrrolizidine alkaloids
RAS	Renin–angiotensin system
RSV	Respiratory syncytial virus
RT	Reverse transcriptase
RA	Rheumatoid arthritis
SCF	Scientific Committee on Food
SAD	Seasonal affective disorder
SLs	Sesquiterpene lactones
SPME GC–MS	Solid-phase microextraction gas chromatography-mass spectrometry

TCM	Traditional Chinese Medicine
TBARS	Thiobarbituric acid reactive substance
UTI	Urinary tract infections
UTIs	Urinary tract infections

ABSTRACT

Medicinal plants ethnobotany is significant in understanding the way of life and customs of individuals. This volume reports different therapeutic plant species, most ordinarily utilized by native persons throughout the planet. The botanical term, family name, dialect name, partly utilized, and the use of the plants have been mentioned in this paper. Medicinal plants are not just a significant resource base for the conventional medication and traditional industry, yet additionally, give work and health security to a huge section of the populace. Around 1178 types of medicinal plants are believed to be traded worldwide, of which 242 species have annual utilization levels of more than 100 metric tons/year. The subsistence demand for medicinal plants has been assessed at 1,95,000 MT per year and the export interest of therapeutic plants has been evaluated at 1,34,500 MT per year. Medicinal plants are the "spine" of plant medicine, which implies more than 3.3 billion individuals in the less developed nations use therapeutic plants consistently. There are almost 2000 ethnic groups on the planet, and pretty much every group has its customary medical information and encounters. This book presents a rundown of restorative plants generally used for relieving disorders, their bioactive elements, and their means of operation.

PREFACE

Therapeutic plants have been used against different infections for millennia, and 80% of the global populace actually relies on home grown medications. The therapeutic plants *Artemisia nilagirica*, *Vaccinium microcarpum*, *Curcuma longa*, *Senegalia visco* and *Hypericum perforatum* are some of which have been successfully utilized against various infections. This natural resource has liberally given us bioactive atoms, for example, hypericin, hydroquinone, epicatechin, artemetin, lupeol and curcumin, which have been utilized against disease break out.

Prior to artificial medications, man depended on the recuperating properties of restorative plants. Certain individuals esteem these plants because of the antiquated conviction which says plants are made to supply man with food, clinical treatment, and other health impacts. It is estimated that around 80% of the 5.2 billion individuals of the world live in the less developed nations, and the World Health Organization gauges that around 80% of these individuals depend only on traditional medication for their essential healthcare needs. Restorative plants are the "spine" of traditional medication, which implies more than 3.3 billion individuals in the less develped nations use therapeutic plants consistently. There are almost 2000 ethnic communities on the planet, and pretty much every community has its own customary medicinal information.

The world is home to different native clans with a rich legacy of information on the employment of therapeutic plants. There are diverse environments and topographical areas that have caused a wide variety of restorative plant species, to such an extent that every clan has its own medicinal plants and customs.

This volume examines the restorative plants with most remedial use. Therapeutic plants have for quite some time been used in customary medication and overall ethnomedicine. This book presents a brief look at the current status of and future patterns in therapeutic plant genomics, advancement, and phylogeny. These powerful fields are at the convergence of phytochemistry and plant science and are concerned about advancement components and systematics of therapeutic plant genomes, beginning and development of plant genotype and metabolic aggregate, cooperation between restorative plant genomes and climate, and connection between's genomic variety and metabolite variety, and so on. The employments of the arising top of the line genomic advancements can be extended from crop plants to conventional restorative plants to assist the therapeutic plant reproduction process, and convert them to the living production line of therapeutic mixtures.

Therapeutic plants have consistently assumed a critical part as hotspots for drug based compounds. Early people, driven by their intuition, taste, and experience, treated their diseases by utilizing plants; henceforth, the historical backdrop of restorative plants is the length of the historical backdrop of people. One test that the development of restorative plants faces is their order. Over the long term, taxonomists accompanied a wide range of approaches for plant classifications like morphologic grouping, anatomic characterization, and chemotaxonomic order. The initial two techniques are assembled under customary groupings, while the third strategy is an advanced way to deal with structuring the plants. The idea of grouping of therapeutic plants based on medicinal use isn't new, however after some time, it has encountered a greater number of constraints. The volume dicusses some of the scientific categorization of restorative plants and how data on their phytochemicals can be utilized for their grouping, issues related with order, and valuable applications.

A

Abelmoschus Esculentus

Abelmoschus esculentus, often known as ladies' fingers or okra, okro or ochro in many English-speaking countries, is a blooming plant in the mallow family. Its delicious green seed pods are prized. Minerals, vitamins, antioxidants, and fiber are all abundant in it. Okra's antecedents are debated, with advocates claiming West African, Ethiopian, and South Asian roots. The plant is grown in tropical, subtropical, and warm temperate climates all over the world, and it's a popular ingredient in Southern gastronomy, as well as Middle Eastern and Indian cuisines (Zhou et al., 2014).

The word okra (sometimes spelled okro or ochro) was originally used in 1679 in the Colony of Virginia, and it was derived from the Igbo word kr. Gumbo is derived from perhaps the Umbundu term ochinggômbo or the Kimbundu word ki-ngombo, and was first reported in American colloquial language around 1805, stemming from Louisiana Creole. The vegetable is known as quimbombó in Cuba and Puerto Rico, and it's used in meals like quimbombó guisado (stewed okra), which is comparable to Southern gumbo.

Regardless of the reality that the name gumbo is commonly used to refer to the dish gumbo in much of the United States, many communities in the Deep South continue to use it to describe to the pods and plants, as well as numerous other versions of the word prevalent throughout the African diaspora in the Americas.

Food and Usage

When the usually consumed parts of plants are heated, the plant's contents are mucilaginous, bringing in the characteristic "goo" or slime; the mucilage includes dietary fibers. Cooking okra with an acidic meal, such as tomatoes, can help to de-slime it by making the mucilage less sticky. Cooked, pickled, raw, or in salads, pods are used in a variety of ways. In underdeveloped nations, okra may be utilized to combat malnutrition and food insecurity.

Okra, despite being categorized as a fruit, has long been used as a vegetable, both on its own and in combination with other foods and materials. It comes in a variety of shapes and sizes when cooked. Slicing and boiling okra in water produces a sticky liquid that thickens the soups and stews.

Medical uses

The various uses of okra are still being researched. In one study, the effect of okra extract on gestational diabetes mellitus in rats was investigated, as well as the possible molecular mechanism. It was discovered to help pregnant rats with diabetes control their blood sugar levels (Yi et al., 2016). Okra is also high in fiber, with approximately 8 medium-sized pods containing 3 grams. Okra boosts fiber intake, improves glycemic management, and enhances insulin sensitivity, according to another study.

Dosage

There is no clinical evidence to support dosage. Tinctures of powdered musk okra seeds in ethanol have been used.

Extracts

Okra extract (Abelmoschus esculentus) is rich in vitamins & minerals and is loaded with antioxidants. It helps support the pancreas, heart, bones, and colon.

Side effects.

Some people may be harmed by eating too much okra.

Okra includes fructans, a type of carbohydrate that can cause gastrointestinal issues. In persons who already have digestive issues, fructans might produce diarrhea, flatulence, cramps, and bloating.

Okra is high in oxalates, which can cause kidney stones. Calcium oxalate is the most frequent form of kidney stone. Eating a diet rich in oxalates, such as okra and spinach, may raise the risk of kidney stones in persons who have formerly had them (Van Jaarsveld, 2013).

Inflammation: Okra contains solanine, a poisonous chemical that in some people can cause joint discomfort, arthritis, and long-term inflammation. Solanine can also be found in potatoes, tomatoes, eggplant, blueberries, and artichokes.

Abrus precatorius

Abrus precatorius, often referred to as jequirity bean or rosary pea, is a Fabaceae bean family herbaceous flowering plant. It's a thin perennial vine with long, pinnately leafed leaves that wraps itself over trees, shrubs, and hedges.

The plant is primarily known for its seeds, which are toxic due to the presence of abrin and are used as beads and musical instruments. Adults and children alike can be killed by ingesting a single seed that has been thoroughly digested. Asia and Australia are the native habitats of this plant. In places it has been introduced, it has a propensity to become weedy and invasive.

Figure 1: Abrus Precatorius.

Source: http://www.plantsoftheworldonline.org/taxon/urn:lsid:ipni.org:names:469605-1

Medicinal Uses.

Because of their beautiful hue, *Abrus precatorius* seeds are highly prized in local jewelry. The majority of beans are black and red in appearance, resembling ladybugs, but various colors are available. Making jewelry from jequirity seeds is rumored to be dangerous (Umadevi et al., 2012). There are recurring stories that workers who puncture the seeds to thread them can become poisoned or even die as a result of a pinprick, however, there appears to be no evidence to back up these claims. A search of the internet turned up 265 scientific papers mentioning *Abrus precatorius*, but none of them mentioned occupational toxicity.

Medical Uses

Abrus precatorius, also known as "Gulaganji" in Kannada, "Kunni kuru" in Malayalam, "kundu mani" in Tamil, "Gurugvinda ginja" in Telugu, and "Gurugvinda ginja" in Telugu, has been utilized in Siddha medicine for millennia. The white variety is used to make oil that is said to have aphrodisiac properties. The leaves are used to make a tea that is intended to treat fevers, coughs, and colds. Because seeds are harmful, they are only taken after being heated. The Tamil Siddhars were aware of plant toxicity and proposed several ways of purification known as "suththi seythal." This is accomplished by boiling and drying the seeds in milk. When exposed to high temperatures, the protein toxin, like castor oil, is denatured, rendering it innocuous.

Dosage

Medicinal Parts	Therapeutic Dosage (for adults)
Detoxified Rosary Pea Powder	30 to 125 mg
Rosary Leaf Powder	500 to 2 grams
Rosary Root Powder	500 to 2 grams

Figure 2: Rosary Pea Dosage.

Source: https://www.ayurtimes.com/rosary-pea-abrus-precatorius/

Extracts

According to the 1889 book 'The Useful Native Plants of Australia,' "In India, the roots of this plant are employed as an alternative for liquorice, even though they are bitter. The roots are regarded as demulcent in Java (Thakur, 2011). The leaves are used to treat swellings when blended with honey and are used as a tea alternative in Jamaica. The seeds have recently been used in cases of ophthalmia under the name "Jequirity," a use that they have long had in India and Brazil."

The herb is also supposed to increase hair development and is utilized in Ayurveda. It's occasionally found in Indian hair care products.

Acorus Calamus

Acorus calamus (also known as sweet flag, sway, and muskrat root, among other terms) is a flowering plant that contains hallucinogenic compounds. It's a long wetland monocot in the Acoraceae family, belonging to the genus Acorus. Although calamus has been used for millennia in traditional medicine to cure digestive issues and discomfort, there is no clinical proof of its effectiveness and safety and swallowed calamus may be poisonous, resulting to its commercial restriction in the United States.

Medical Uses

External uses: Its powder is sprinkled over burn injuries and cuts as a blood cleanser, hemostatic, vasoconstriction, and healing action. Tincture of bark is used for uttara basti (enema through vagina) in leucorrhoea. It's also utilized for rectal prolapse tub baths and as gargles for oral problems.

Internal Uses: Diarrhoea, dysentery, piles, Anthelemintic, bleeding disease, cough, and urinary disorders might all benefit from this herb. Gum is used to treat dysuria and libido loss. The usage of raw legume + sugar powder in premature ejaculation has been discovered to be beneficial.

Dosage

50 -100 ml for decoction

Extracts and other uses

For millennia, *A. calamus* has been a valuable commodity in many cultures. Calamus essential oil is appreciated in the cosmetics industry because of its aroma, which has been used for its healing properties for some maladies such as gastrointestinal problems and pain relief. In Europe, the rhizome's extract is used to flavor dishes, alcoholic beverages, and bitters. It was also used to manufacture confectionery in the past. The dried and ground-up roots have a compound that is effective as an insecticide (Stanisavljević et al., 2007).

Acacia Arabica

Figure 3: Acacia Arabica.

Source: https://www.dreamstime.com/photos-images/acacia-arabica.html

Synonyms

Acacia arabica, Acacia senegal, Acacia verek, Arbre à Gomme Arabique, Bum Senegal, Bomme Arabique, Bomme de Senegal, Bummae Momosae, Goma Arábiga, Gomme Acacia, Gomme Arabique, Gomme d'Acacia, Gomme Sénégal, Gommier Blanc, Gum Acacia, Gum Arabic, Khadir, Kher, Kumatia, Mimosa senegal, Senegalia senegal.

Medical Uses

High Cholesterol. Acacia does not appear to decrease cholesterol levels when taken orally.

Plaque on the teeth. According to preliminary research, chewing acacia gum for 10 minutes five times per day for seven days reduces dental plaque more effectively than sugar-free chewing gum (Seth, 2003).

Loss of weight. Early research suggests that ingesting 30 grams of powdered acacia daily may aid weight loss.

Side Effects

When eaten by mouth in proportions usually seen in meals, acacia is LIKELY SAFE for most adults.

When taken orally in doses utilized for medical purposes, acacia is POSSIBLY SAFE. For the past six weeks, doses of up to 30 grams per day have been utilized successfully. It can, however, cause minor side effects such as gas, bloating, nausea, and loose stools.

Dosage

The right amount of acacia hinges on a variety of factors, including the user's age, health, and other circumstances. There is currently insufficient scientific data to define a suitable dosing range for acacia. Please remember that natural products aren't always safe and that dosages are crucial. Before using, be sure to read the product label and contact your pharmacist, physician, or another healthcare expert.

Acacia Catechu

Other Name(s)

Black Catechu: Acacia à Cachou, Acacia catechu, Acacia Catechu Heartwood Extract, Black Cutch, Cachou de Pegu, Cachou Noir, Cachou, Cashou, Catechu nigrum, Catecu, Cutch, Dark Catechu, Er Cha, Khair, Khadira, Mimosa catechu, Pegu Catechu.

Medical Uses

Catechu is a type of herb. Medicine is made from leaves, branches, and wood. Although the two forms of catechu, black catechu and pale catechu, contain slightly different ingredients, they are both used for the same purposes and in the same dose (Saslis-Lagoudakis et al., 2014).

Catechu is prescribed for diarrhea, nasal and throat inflammation, dysentery, colon swelling (colitis), hemorrhaging, indigestion, osteoarthritis, and cancer.

Catechu is used to treat skin problems, hemorrhoids, and physical trauma, as well as to cease bleeding and dress cuts.

Catechu is used in mouthwashes and used to treat gum disease (gingivitis), mouth pain and swelling (stomatitis), sore throat, and mouth ulcers.

Side Effects

Catechu is safe in dietary quantities. However, there isn't enough data to tell if it's safe in the larger doses used in medicine. Flavocoxid (Limbrel, Primus Pharmaceuticals), a combination medication including catechu, was found to be safe in research investigations lasting up to 12 weeks. However, some persons may get liver problems as a result of this combo product. This adverse effect does not appear to be widespread, and it may only be experienced by those who have an allergic response to it.

It's also unclear whether applying catechu straight to the skin is safe.

Dosage

The right dose of catechu is determined by a variety of factors, including the user's age, health, and other circumstances. There is currently insufficient scientific data to define an appropriate dosing range for catechu. Keep in mind that natural products aren't always safe and that dosages are crucial. Before using, be sure to read the product label and contact your pharmacist, physician, or another healthcare expert.

Adhatoda Vasica

Other Names

Justicia adhatoda, commonly known in English as Malabar nut, adulsa, adhatoda, vasa, vasaka is a medicinal plant native to Asia, widely used in traditional medicine.

Figure 4: Adhatoda Vasica (Vaska).

Source: https://theindianmed.com/adhatoda-vasica-vasaka-uses-and-benefits/

Medical Uses

- For thousands of years, Adhatoda Vasica has been utilized in traditional Indian medicine to treat respiratory problems. Bronchitis, TB, and other lung and bronchiole illnesses can all be helped with (Nazim et al., 2020).
- The leaves can be decocted and used as a herbal therapy for cough and other cold symptoms. The relaxing action relieves throat irritation, while the expectorant action helps clear phlegm accumulation in the airway, making adhatoda a good sore throat cure.
- Internal and external bleeding, such as peptic ulcers, piles, and bleeding gums, have all been treated with Adhatoda Vasica.
- Antispasmodic, expectorant, and blood cleansing properties are all present in this herb.
- Adhatoda Vasica has also been used to hasten the delivery process.

Adhatoda is deemed safe when used and dosed as directed. This herb's safety in children has not been established, thus users should be taken unless advised by a medical professional. This product is not indicated for use during pregnancy (unless at the time of birth, and then only under the supervision of a physician.) When taking this plant with other medications or supplements that have expectorant or antispasmodic properties, precautions should be taken.

Dosage

In doses of 15 to 30, the medicine vasaka is commonly taken in the form of juice produced from its leaves, combined with ginger or honey. The dried leaves can be administered in powder form in doses of 2 grams, or they can be prepared into a decoction. Both the decoction and powder are used in Ayurvedic medicine to treat a variety of respiratory ailments.

The bark and root have the same therapeutic properties as the leaves. The powdered root bark is administered in 0.75 to 2 grams and a decoction of the bark is administered in 30 to 60 ml doses.

Allium Cepa

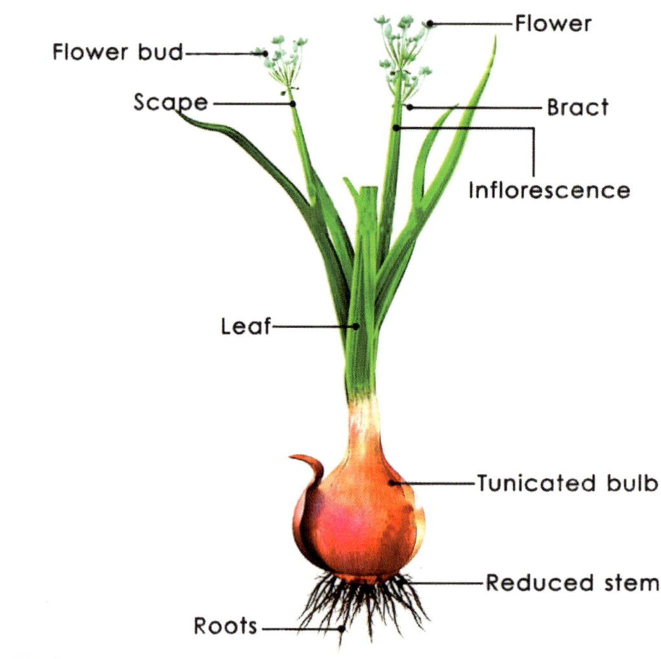

Figure 5: The Onion.

Source: https://www.istockphoto.com/photo/allium-strain-gm531345335-55223648

Other Names

The onion (*Allium cepa* L., from Latin cepa "onion") is a plant that is the most frequently grown member of the genus Allium. It is also known as the bulb onion or common onion. The shallot is an onion botanical variation. The shallot was classed as a distinct species until 2010.

Uses

Scarring. The majority of study demonstrates that applying an onion extract ointment to the skin for at least 10 weeks, alone or in combination with other components, improves the look of scars. Using onion extract with other components for a shorter period does not appear to work (Tuttolomondo et al., 2014).

It has also been proven to various extents to work on several issues including:

- High blood pressure.
- Insomnia.
- Prostate cancer.
- Stretch marks.
- Asthma.
- Bronchitis.
- Colds.
- Cough.
- Fever.
- Loss of appetite.
- Preventing hardening of the arteries (atherosclerosis).
- Swelling (inflammation) of the mouth and throat.
- Upset stomach.
- Wounds.
- Other conditions.

Side Effects

When consumed by mouth, onions are LIKELY SAFE when consumed at levels that are regularly seen in meals. When ingested orally as a medicine, onion extract is POSSIBLY SAFE for most people. For 12 weeks, daily doses of up to 900 mg of onion extract were used safely. After eating onion, you may have stomach discomfort or agony.

When it comes to applying onion extract to the skin, it is VERY LIKELY SAFE. Following skin contact with onion, side effects may include irritation or dermatitis.

Dosing

Products containing onion extract and other ingredients such as heparin plus allantoin (Contractubex, Merz Pharmaceuticals), allantoin plus pentaglycan (Kaloidon gel, Laboratori Farmacologici Milanesi), or a silicon derivative (Cybele scagel, Bangkok Botanica) have been applied to the affected area for 10 weeks to 6 months to treat scarring work (Teixidor-Toneu et al., 2018).

Allium Sativum

Synonyms

Garlic (*Allium sativum*) is a bulbous flowering plant species of the onion genus Allium. It is related to the onion, shallot, leek, chive, Welsh onion, and Chinese onion, among others

Figure 6: Garlic.

Source: https://www.jessicagavin.com/benefits-of-garlic/

Medical Uses

Garlic has been used to treat artery hardening (atherosclerosis) and high blood pressure (hypertension). If you have cardiovascular disease or increased blood pressure, speak with your doctor for more details. Garlic has also been used to keep ticks at bay. Some herbal/dietary supplement products have been discovered to have potentially hazardous impurities/ additives. More information about the brand you take can be obtained from your pharmacist.

Side Effects

Bad breath and body odor, stomachache, and heartburn are all possible. Contact your doctor or pharmacist right away if any of these side effects continue or worsen. Inform your doctor right away if any of the following rare but serious side effects occur: easy or unusual bruising/bleeding. It is uncommon for this product to cause a severe adverse reaction (Silva et al., 2010). However, if you detect any of the following symptoms of severe allergic response, get emergency medical attention: Rashes, itching/ swelling (particularly of the face, tongue, and neck), severe disorientation, and difficulty breathing.

Extracts

An alcoholic extract of the bulb or whole garlic plant *Allium sativum* (Liliaceae) with anticancer action. Garlic has long been seen to be a helpful plant for health, having been used as an antihelmintic, rubefacient, anti-infective, and antihypertensive. Garlic extracts, whether fresh or aged, contain chemicals such as diallyl and allyl propyl disulfides, which have significant antioxidant and cholesterol-lowering qualities; regular consumption may help prevent atherosclerosis and cardiovascular disease.

Aloe Vera

Medical Uses and Extracts

Aloe vera is a succulent species of plant of the Aloe genus. With over 500 kinds, Aloe is extensively dispersed and regarded as an invasive species in many parts of the world.

It is an evergreen perennial that originated in the Arabian Peninsula but now grows in tropical, semi-tropical, and dry areas all over the world. It is grown for commercial purposes, mostly as a topical therapy that has been used for centuries. The species is appealing as a decorative plant and thrives indoors as a potted plant (Sepúlveda-Cuellar et al., 2021).

It is found in a variety of consumer products, including beverages, skin lotion, cosmetics, ointments, and gel for mild burns and sunburns.

Commercial goods are made from two Aloe vera substances: a clear gel and its yellow latex. Topical treatments containing aloe gel are commonly used to treat skin diseases such as burns, wounds, frostbite, rashes, psoriasis, cold sores, or dry skin. Aloe latex is used separately or in a product with additional substances that is swallowed for constipation treatment. Aloe latex is available in two forms: dried resin and "aloe dried juice."

Dosage

The California Office of Environmental Health Hazard Assessment classified non-decolorized aloe vera leaf extract, along with goldenseal, as "chemicals known to the state to cause cancer or reproductive harm."

The use of topical aloe vera has not been linked to any severe negative effects. Oral intake of aloe vera is potentially toxic, causing abdominal pains and diarrhea, which might reduce medication absorption.

Side Effects (Used in long-term and high doses)

- blood sugar (hypoglycemia)
- Burning and itching of the skin
- Stomach pain and cramps
- Diarrhea, kidney problems, blood in the urine, low potassium, muscle weakness, weight loss, and heart disturbances
- Liver problems

Andrographis Paniculata

Synonyms and usage

Andrographis paniculata, often known as creat or green chiretta, is a yearly herbaceous plant indigenous to India and Sri Lanka in the Acanthaceae family (Salim et al., 2019).

It is abundantly grown in Southern and Southeastern Asia, an area that has traditionally been used to cure bacterial infections and various ailments.

For the most part, the leaves and roots were used for this purpose. In certain circumstances, the entire plant is used.

Figure 7: Stone Age *Andrographis Paniculata* Leaves.

Source: https://www.indiamart.com/proddetail/andrographis-paniculata-leaves-9513266412.html

Medical Uses and Effectiveness

Possibly useful for common colds. Taking Andrographis by mouth, either alone or as part of a herbal combination product, appears to alleviate cough and sore throat symptoms in some persons suffering from the common cold. It's unclear whether andrographis can help avoid colds.

Osteoarthritis. Taking andrographis extract by mouth regularly appears to improve pain and stiffness in patients with mild to moderate knee osteoarthritis.

Infection of the tonsils and throat (tonsillopharyngitis). To relieve fever and pain from tonsillitis, taking andrographis orally daily duties roughly as well as acetaminophen (Tylenol).

Dosage and Extracts

Adults have most commonly taken andrographis in doses ranging from 90-600 mg daily for up to 12 weeks. It is also available in a variety of combo items. Andrographis extracts are often regulated by the amount of a substance known as andrographolide that they carry. Normally, this spans from 2% to 50%. Speak with a healthcare expert to determine what type of product and dose might be most appropriate for a specific ailment (Schultz et al., 2020).

Andropogon annulatus

Synonyms

Common names for *Andropogon annulatus* include marvel grass, Diaz bluestem, Kleberg bluestem, Hindi grass, ringed dichanthium, shade grass, medio bluestem (var. papillosum), jargu grass, Delhi grass, vuda bluegrass, two-flowered golden-beard, Santa Barbara grass

Medicinal uses.

Poaceae (the family of *Andropogon annulatus*) is represented in Pakistan by 158 genera and 492 species and is extremely important economically and medicinally. Some Poaceae plants are used as antihypertensive, antidiabetic, anti-inflammatory, anthelmintic, antiulcer, diuretic, and antioxidant medications.

Dosage

Not much is known about the dosage but being a forrage plant it is made into a tea or ticture and then consume.

Annona Squamosa

Synonyms

Sugar Apple, Sweetsop, Custard Apple

Figure 8: Custard Apple.

Source: https://www.1mg.com/ayurveda/custard-apple-124

Medicinal Uses and Extracts

Medicinal properties have been recorded for the leaves, shoots, bark, and roots. They are all medicinal and used to treat diarrhoea and dysentery. Vermicidal qualities are found in green fruits, seeds, and leaves. In the West Indies, young shoots mixed with peppermint are used to treat colds and chills. The leaves are used to decrease uric acid levels in the blood in Cuba. The unripe fruit has astringent properties. The root has a strong purgative effect (Saslis-Lagoudakis et al., 2011). The bark and leaves of this plant, along with those of *Annona muricata*, are used to make a sedative infusion. An infusion of the leaves and berries is used to alleviate rheumatism and help digestion. Sleeplessness is treated with oil distilled from the leaves rubbed on the head. The seed can be powdered down and used as a vermifuge and extracts can also be used as an anticancer medicine

Dosage and Side Effects

The recommended dosage of custard apple is custard apple powder - 1-2 teaspoons or as per your requirement. Preparations using Custard apple seeds should be avoided. External application of Custard apple seed powder has been linked to incidences of acute pain, redness, and eye damage, according to research.

Arachis Hypogaea

Synonym

The peanut, also known as the groundnut, goober (in the United States), pindar (in the United States), or monkey nut (in the United Kingdom), and taxonomically designated as Arachis hypogaea, is a legume crop farmed primarily for its edible seeds. It is commonly cultivated throughout the tropics and subtropics, and it is significant to both small and major industry players. It is classed as a grain legume[4] as well as an oil crop due to its high oil content.

Medicinal Uses

The seed oil is aperient, demulcent, emollient, and pectoral. The seed is mostly utilized as a nutritious food. In traditional medicine, the seeds have been utilized as an anti-inflammatory, aphrodisiac, and decoagulant. Peanuts are mentioned in a few folk pharmacopeias. In China, the nuts are regarded as demulcent, pectoral, and peptic; the oil is aperient and emollient and is used internally in milk to treat gonorrhea and externally to treat rheumatism. Peanut is used in folk treatments for plantar warts in Zimbabwe (Reinaldo et al., 2020).

Side Effects

The possibility of teratogenic, carcinogenic aflatoxins contaminating damaged or rotting seeds is of the utmost concern. The aflatoxin-generating moulds produce two major poisons, aflatoxins B and G, as well as their less deadly dihydro counterparts, aflatoxins B2 and G2 (Aspergillus flavus et al). Because there is no effective technique to eliminate poisons from feed and meals, prevention of mould formation is the cornerstone (however, peanut oils are free of aflatoxins because of alkaline processing). If there is any indication of an allergy, avoid it.

Extracts

The extract is commonly utilized in medicinal and cosmetic formulations as a carrier. It is made up of a range of ingredients. Amino acids, fatty acids such as oleic acid, linoleic acid, and arachidic acid, to mention a few, vitamins, minerals, flavonoids, carbohydrates, and a variety of other inorganic components It contains various fatty oils and essential oils, thus it gives skin-soothing effects. It is quickly absorbed in the skin, and it also aids another element in exerting an effect on the skin, acting synergistically.

Abscess Root

Synonyms

American Greek Valerian, Blue Bells, False Jacob's Ladder, Polémoine Rampante, Polémonie Rampante, Polemonium reptans, Sweatroot, Valeriana Griega.

Medicinal Uses

Abscess root is a type of plant. The root is powdered and used to make medicinal tea.

Abscess root is used to treat fever, edema, and inflammation, as well as to induce perspiration. It is also used as a cleansing agent (an astringent) and to release chest congestion in order for it to be coughed up (as an expectorant) (Rawat et al., 2020).

Abscess root is also known as "false Jacob's ladder" due to its astringent action, which is comparable to that of Jacob's ladder. These two plants are used in similar ways but have different chemical properties. Make sure not to mix them.

Side Effects

There isn't enough data to say whether abscess root is safe. It may cause stomach trouble and sneeze as adverse effects.

Dosing

The optimal dose of abscess root is determined by numerous factors, including the user's age, health, and other conditions. There is currently insufficient scientific evidence to define an acceptable range of dosage for abscess root.

Antipyretic

An antipyretic is a drug that lowers fever. Antipyretics cause the hypothalamus to suppress a temperature increase caused by prostaglandins. The body then works to lower the temperature, resulting in a fever reduction.

Most antipyretic medicines serve additional functions. In the United States, the most commonly used antipyretics are ibuprofen and aspirin, which are nonsteroidal anti-inflammatory drugs (NSAIDs) used primarily as analgesics (pain relievers) but also have antipyretic properties; and paracetamol (acetaminophen), an analgesic with weak anti-inflammatory properties (Peng et al., 2017).

Astringents

Astringents (sometimes known as astringents) are chemicals that shrink or constrict bodily tissues. The term comes from the Latin adstringere, which means "to bind tightly." Astringents include calamine lotion, witch hazel, and the Californian plant yerba mansa.

Astringency, or the dry, puckering sensation generated by tannins in unripe fruits, allows the fruit to ripen by discouraging consumption. Ripe fruits and fruit parts such as blackthorn (sloe berries), Aronia chokeberry, chokecherry, bird cherry, rhubarb, quince and persimmon fruits, and banana skins are extremely astringent; citrus fruits such as lemons are only somewhat astringent. Tannins, a type of polyphenol, attach to salivary proteins and cause them to precipitate and aggregate, producing a rough, "sandpapery", or dry sensation in the mouth.

Adam's Needles

Synonyms

Yucca filamentosa, Adam's needle and thread, is a species of flowering plant in the family Asparagaceae native to the southeastern United States.

Medicinal Use

- Inadequate Evidence for
- Cholesterol.
- Arthritis.
- Blood pressure

- Migraines.
- Digestive problems.
- Diabetes.
- Inadequate blood circulation
- Skin issues.
- Other circumstances.
- More research is needed to determine the efficacy of yucca for these applications.

Figure 9: Yucca filamentosa.

Source: https://en.wikipedia.org/wiki/Yucca_filamentosa

Side Effects

When consumed in the amounts seen in foods, yucca is LIKELY SAFE. When consumed by mouth for a short period, Yucca is POSSIBLY SAFE. Some people may get stomach trouble as a result of it.

There isn't enough information about the long-term safety of taking yucca or applying it to the skin. Yucca can cause allergic responses in certain people. Skin rash and trouble breathing are possible symptoms.

Dosage

The proper yucca dose is determined by numerous factors, including the user's age, health, and other factors. There is currently insufficient scientific evidence to identify an appropriate range of yucca dosages. Please remember that natural products are not always safe, and dosages can be critical. Before using, be sure to read the product label and consult your pharmacist, physician, or other healthcare expert (Ningthoujamet et al., 2012).

Acacia Nilotica

Synonyms

Acacia nilotica or *Vachellia nilotica* (also known as gum arabic tree, babul, thorn mimosa, Egyptian acacia, or thorny acacia) is a Fabaceae flowering plant tree. Its native range includes Africa, the Middle East, and the Indian subcontinent. It is also a Weed of National Significance in Australia (a large invasive species in Australia) and a Federal Noxious Weed in the United States.

Medicinal Use and Extracts

Acacia Nilotica has a wide range of therapeutic applications. Acacia Nilotica is commonly utilized in traditional medicine. This plant contains antimicrobial, anti-plasmodial, and antioxidant properties and is used to treat HIV, hepatitis C virus, and cancer. It can be used to treat sexual diseases, nausea, burns and wounds, stomach aches, and diarrhea (Nadda et al., 2020). Boiled leaf extract is used to treat chest pain or pneumonia, weakness and chest ailments, fever, malaria, headache, coughing, sore joints, backache, and stomach ulcers, among other things. Acacia Nilotica fresh pods are useful in sexual problems such as spermatorrhoea, sperm viscidity loss, and recurring night discharges. *Acacia nilotica* can help prevent premature ejaculation, alleviate discomfort in acute gonorrhea and leucorrhoea, and treat urine-genital disorders such pelvic prolapse.

Dosage and Side effects.

Acacia is LIKELY SAFE for most adults when consumed at proportions found in food.

Acacia is POSSIBLY SAFE when taken by mouth in medical quantities. For 6 weeks, doses of up to 30 grams per day were taken safely. It can, however, cause minor side effects such as gas, bloating, nausea, and loose stools.

The optimal dose of acacia is determined by numerous factors, including the user's age, health, and other conditions. At this time, there is insufficient scientific evidence to define a suitable dosing range for acacia. Keep in mind that natural products are not always safe, and dosages can be critical. Before using, be sure to read the product label and consult your pharmacist, physician, or other healthcare experts.

African Caper

Synonyms

Capparis tomentosa, the woolly caper bush or African caper, is a plant in the Capparaceae family and is native to southern Africa.

Capparis tomentosa favors deciduous bushlands and forest savannas, where it grows on ant hills. Because Lepidoptera species (moths and butterflies) rely on a limited variety of plants throughout their larval stages, the plant is a vital food source for them (Miara et al., 2018). The leaves are also eaten by wildlife, however, they have been linked to toxicity in animals. Humans and monkeys consume the fruits. Because the flowers attract bees, this species is also useful for honey production.

Figure 10: Capparis tomentosa.

Source: http://www.plantsoftheworldonline.org/taxon/urn:lsid:ipni.org:names:146824-1

Medical Uses and extracts

Many parts of the plant are considered to be toxic but it is a plant that has been used in traditional medicine for millennia. The root powder is used to treat rheumatism, colds coughs and tuberculosis. It is also used for stomach upsets rectal prolapse and a number of STIs.

The powder has also been used to treat open wounds, so far as snake bites, skin conditions and hemorrhoids.

Dosage and Side effects

Capers are LIKELY SAFE for most individuals when consumed as food when taken by mouth. When taken by mouth as a medicine, caper fruit extract is POSSIBLY SAFE.

When administered to the skin, there is insufficient evidence to determine if capers are safe. Some people may experience negative effects such as skin rash and discomfort.

The right dose of capers to use as a medication is determined by several factors, including the user's age, health, and other circumstances. There is currently insufficient scientific data to define an appropriate dosing range for capers.

Acyl-CoA

Acyl-CoA is a family of coenzymes involved in the metabolism of fatty acids. Acyl-CoAs are vulnerable to beta-oxidation, which results in the formation of acetyl-CoA. The acetyl-CoA enters the citric acid cycle and produces many ATP equivalents. Fats are transformed to ATP, the universal biological energy carrier, in this way.

Acyltrans

Any of several transferase enzymes that catalyze the transfer of acyl groups between lipids

Acyltransferase is a type of transferase enzyme that acts upon acyl groups. Examples include Glyceronephosphate O-acyltransferase, Lecithin-cholesterol acyltransferase, Long-chain-alcohol O-fatty-acyltransferase (Menendez-Baceta et al., 2015).

Angiotensin Converting Enzyme

The renin–angiotensin system (RAS) regulates blood pressure by regulating the volume of fluids in the body. Angiotensin-converting enzyme, or ACE, is a key component of the RAS. It transforms angiotensin I, a hormone, into angiotensin II, an active vasoconstrictor. As a result, ACE indirectly raises blood pressure by constricting blood vessels. ACE inhibitors are a type of medicine that is commonly used to treat cardiovascular disorders.

Leonard T. Skeggs Jr. identified the enzyme in 1956. In the year 2002, the first crystal structure of human testis ACE was solved. It is most prevalent in pulmonary capillaries, although it can also be found in endothelium and renal epithelial cells.

Antiacetylcholinesterase

The major cholinesterase in the body is acetylcholinesterase, commonly known as AChE or acetylhydrolase. It's a neurotransmitter-degrading enzyme that catalyzes the degradation of acetylcholine and other choline esters. AChE is primarily located at neuromuscular junctions and cholinergic chemical synapses, where its functioning acts to terminate synaptic transmission. It is a member of the carboxylesterase enzyme family. Organophosphorus chemicals, such as nerve agents and insecticides, are the principal inhibitors of this enzyme.

Aryl Hydrocarbon Hydroxylase

Aryl hydrocarbon hydroxylase (AHH) is a multicomponent, microsomal-bound enzyme system that transforms a wide range of lipid-soluble chemicals into water-soluble forms for elimination. Endogenous and exogenous substances such as steroid hormones, barbiturates, pesticides, polycyclic aromatic hydrocarbons (PAH), and entire cigarette smoke can all induce the enzyme system (Muchugi et al., 2006). Recent findings show that inducibility is regulated by the host gene and that inducibility of this enzyme correlates with carcinogenic sensitivity to PAH in animals and bronchogenic squamous cell carcinoma in humans (probably cigarette smoke-induced).

Alpha-linolenic Acid

Alpha-linolenic acid (ALA) is an omega-3 fatty acid that can be found in nuts like walnuts. It is required for appropriate human development and growth.

By assisting in the maintenance of proper heart rhythm and pumping, alpha-linolenic acid is thought to lower the risk of heart disease. It may also help to prevent blood clots. Vegetable oils like flaxseed and canola oil, as well as red meat and dairy products, are common dietary sources.

The most prevalent uses for alpha-linolenic acid include heart and blood vessel illnesses like artery hardening, heart disease, and high blood pressure. It's also utilized for other ailments, although most of them aren't backed up by scientific data.

Adenosine Monophosphate

Adenosine monophosphate (AMP) is a nucleotide that is also known as 5'-adenylic acid. It is an ester of phosphoric acid and the nucleoside adenosine, and it has a phosphate group, the sugar ribose, and the nucleobase adenine. It takes the form of the prefix adenylyl- as a substituent.

Interconversion of AMP to ADP and/or ATP is vital in many cellular metabolic activities. AMP is also used in the production of RNA. AMP is found in all known life forms.

B

Bamboo

Synonyms

Arrow Bamboo, Arundinaria japonica, Bambou, Bambou Flèche, Bambou du Japon, Bambou Métaké, Bambú, Pseudosasa japonica, Sasa japonica, Yadake.

Medicinal Uses

It has been commonly used to treat Asthma.

Cough.

Gallbladder problems.

Other conditions.

Side Effects

There isn't enough reliable information available about bamboo to know if it is safe.

-Dosage

The right amount of bamboo to take depends on several factors, including the user's age, health, and other circumstances (Maruca et al., 2019). There is currently insufficient scientific data to define a suitable dosing range for bamboo.

Extracts

Bamboo shoots are used in traditional Chinese medicine to help with labor and placenta ejection by stimulating uterine contractions. Shoot poultices are commonly used to clean wounds and treat infections. A decoction of bamboo shoots and honey is used to cure respiratory problems.

Bassia Latifolia

Synonyms

Bassia latifolia Roxb. *Bassia longifolia* L. *Madhuca indica* J.F.Gmel. *Madhuca latifolia*, Mahua. In india it is commonly said that, "The flowers are Sukrala (Semen augmentator), Balya (Restoratives), Pittakara (increases pitta), Vatahara (reduces Vata), Ahridya (harmful for the heart)."

Medicinal Uses

Cooling, tonic, and demulcent properties are attributed to the blooms. Coughs, colds, and bronchitis are all treated with them. Leprosy is a disease that is treated using bark. In Nepal, diabetic patients are given a decoction of the bark. Externally, it is used to cure itchy skin and bleeding gums. The oil derived from the seeds is used to treat skin ailments.

Figure 11: File:Bassia latifolia(Latin),Mahua Tree.JPG.

Source: https://commons.wikimedia.org/wiki/File:Bassia_latifolia(Latin),Mahua_Tree.JPG

Doses and Side Effects

The flowers are Sukrala (Semen augmentator), Balya (Restoratives), Pittakara (increases pitta), Vatahara (reduces Vata), Ahridya (harmful for the heart). Cooling, tonic, and demulcent properties are attributed to the blooms. Coughs, colds, and bronchitis are all treated with them. Leprosy is a disease that is treated using bark. In Nepal, diabetic patients are given a decoction of the bark (Martínez & Luján, 2011). Externally, it is used to cure itchy skin and bleeding gums. The oil derived from the seeds is used to treat skin ailments.

Barleria Lupulina

Synonymsm Usage

The hop-headed barleria, Barleria lupulina, is a plant in the Acanthaceae family. It can be found in Southeast Asia.

Medical *Usage*

The leaf juice is administered to halt bleeding when cut, and the leaf paste is used as a poultice to relieve discomfort. It's well-known in Thai traditional medicine for its anti-inflammatory properties against insect bites, snake bites, herpes simplex, herpes zoster, and varicella-zoster virus lesions, as well as its diuretic and anti-amoebic properties. It is also utilized for its medical value, with the leaf juice being used to halt bleeding when cut and the leaf paste being used as a poultice to relieve pain. It has a powerful inhibitory effect on microorganisms that cause acne.

Dosage and Side Effects

In the future, rigorous scientific evaluation, including isolation of relevant phytochemicals, their mechanism of action, toxicity, and adequate standardization, will be researched to convert these pharmacological properties into modern pharmaceuticals.

The envelope or covering that surrounds a plant's roots, trunk, and branches. Non-pharmacologically significant barks are alphabetized under specified names.

Bark

Synonym

The envelope or covering that surrounds a plant's roots, trunk, and branches. Non-pharmacologically significant barks are alphabetized under specified names.

Medical uses and extractions

Fevers, diabetes, renal disorders, sore eyes, and hemorrhaging have all been treated using tree barks, as well as general tonics. There are two recipes for multi-bark medications that can be used for a variety of purposes. Infusions or decoctions of barks are commonly utilized. The drugs are either ingested or used as a wash (Mati & de Boer, 2011).

Usage of the bark has usually been denoted by the accumulation of the active substance and made into powders, tinctures and oils. This fits into the variety of extractions that can be produced. It is dosed in a variety of concentrations and worked effectively.

Dosage and side effects

Due to the concentrations of a variety of plants and concentrations of a variety of active substances in the product.

Berberis aristata, popularly known as Indian barberry, "daru haldi," or tree turmeric, is a Berberidaceae (Berberis family) and Berberis genus shrub. The genus has about 450-500 deciduous evergreen plant species that can be found in temperate and sub-tropical areas of Asia, Europe, and America. *B. aristata* is found in the Indian and Nepalese Himalayas. It can also be known to occur in Sri Lanka's wet zone.

Bixa Orellana

Synonyms and origin

Bixa orellana (Achiote) is a shrub found in northern South America and northern Mexico. Bixa orellana is cultivated in several countries across the world. The tree is well known for producing annatto, a natural orange-red condiment made from the waxy arils that cover its seeds (also known as achiote or bijol). Cochinita pibil, chicken in achiote, caldo de olla, and nacatamal are just a few of the classic recipes made using pulverized seeds in Central and South America, Mexico, and the Caribbean (Li et al., 2015). Annatto and its extracts are also used as an industrial food coloring to give butter, cheese, margarine, ice creams, meats, and sauces a yellow or orange color.

Figure 12: Horticultural Impex Natural Bixa Orellana Seeds.

Source: https://www.indiamart.com/proddetail/bixa-orellana-seeds-10120925162.html

Medical Uses

Bixa orellana is a plant that has been utilized in traditional medicine. The tree has been utilized in Ayurveda, India's traditional medicine, where various components of the plant are regarded to be therapeutic. The astringent characteristics of the Annatto pulp and other portions are used to treat diarrhea, jaundice (leaves), and other renal ailments.

Blumea Balsamifera

Synonyms

Blumea balsamifera is a flowering plant in the Blumea genus of the Asteraceae family. Ngai camphor and sambong are two other names for it (also sembung).

Medical uses

It is mostly used as a diuretic (or "water pill") and to alleviate cold symptoms. Sambong is a diuretic plant that reduces high blood pressure by treating urolithiasis (urinary tract or kidney stones) and urinary tract infections (López-Malo et al., 2020). Sambong is an expectorant, anti-diarrheal, and that treats some of the symptoms of the common cold. It's also sometimes used as a wound astringent.

Extracts and side effects

Sambong is an aromatic shrub, that grows from 1 to 4 meters in height. It is considered as a weed in some countries and is difficult to eradicate.

Brassica Integrifolia

Other Names

Brown mustard, Chinese mustard, Indian mustard, leaf mustard, Oriental mustard, and vegetable mustard are all names for the mustard plant Brassica juncea.

Brown mustard, Chinese mustard, Indian mustard, leaf mustard, Oriental mustard, and vegetable mustard are all names for the mustard plant Brassica juncea.

Figure 13: Brassica juncea.

Source: https://en.wikipedia.org/wiki/Brassica_juncea

Medicinal Uses

Brassica juncea has a range of uses: it produces seed oil, the crushed seed is used to make mustard, and it is used in a variety of vegetable dishes. It's also utilized as a fodder plant and for medicinal purposes.

Dosage and general Use

When ingested in food amounts, Kohlrabi is LIKELY SAFE when taken by mouth. There isn't enough solid data to tell if kohlrabi is effective at medicinal doses or what the potential side effects are.

When used on the skin, there is insufficient credible evidence to determine whether kohlrabi is safe or what the potential adverse effects are.

The sufficient combination of kohlrabi to eat is determined by several factors, including the user's age, health, and other circumstances. There is currently insufficient scientific data to define an optimal dosing range for kohlrabi. Keep in mind that natural products aren't always safe and that dosages are crucial (Kujawska et al., 2017).

Breynia Patens

Synonyms

- Scientific name: Breynia patens (Roxb.) Rolfe
- Sanskrit: Bahupraja, Bahupushpa, Kamboji
- Kannada: Kamboji
- Malayalam: Aattacherukola, Ekdania
- Bengali: Silpati
- Marma: Rownee
- Chakma: Hamussa Gach, Mach Chok Bichi Gach
- Tripura: Satamsakatoma, Taipak
- Khumi: Aghkamchi

Medicinal Use

Breynia patens is used to treat a wide range of conditions, including body discomfort, skin inflammation, hyperglycemia, diarrhea, dysentery, and toothaches. Conjunctivitis is treated with stem juice. The antibacterial poultice of leaves is applied to boils to speed up suppuration or pus formation. The fruits are offered to people who have dysentery. In fits, the roots are useful.

Baby's Breath

Synonym

Gypsophila (/dpsfl/) is a flowering plant genus belonging to the Caryophyllaceae family. Eurasia, Africa, Australia, and the Pacific Islands are their native habitats. Turkey contains a notably diverse Gypsophila taxonomic assemblage, with approximately 35 indigenous species. Some Gypsophila species have been introduced to different areas.

The name gypsophilios comes from the Greek words gypsos ("gypsum") and philios ("loving"), and refers to the gypsum-rich soils on which some species grow. Baby's breath, or babe's breath, is a common term for plants of the genus Gypsophila, which also corresponds to the well-known ornamental species Gypsophila paniculata (Kunwar et al., 2016).

Medicinal Uses

Owing to a little flower present in most flower shops and many gardens, researchers have discovered a breakthrough that has the potential to save thousands of lives each year.

Gypsophila Paniculata – often known as Baby's Breath – molecules were found in testing to break down the membrane of cancer cells. Antibody-based treatments will be able to fight the tumour more effectively as a result of this.

They discovered that an extract from the white bloom can increase the effectiveness of anticancer medications by a million times (1,000,000). The discovery was made by Leukemia Busters researchers.

Side Effects

Although no mention of this species has been found, at least one member of the genus possesses a saponin-rich root. Despite their toxicity, these chemicals are poorly absorbed by the body and hence tend to pass through unharmed. Heat also breaks them down, so a long, slow bake can destroy them.

Bael, Bengal Quince

Synonyms

Aegle marmelos, sometimes called bael, golden apple, Japanese bitter orange, stone apple, or wood apple, is an exotic tree endemic to the Indian subcontinent and Southeast Asia. As an established species, it can be found in India, Bangladesh, Sri Lanka, and Nepal. Hindus revere the tree as a sacred site.

Medicinal Use and Extracts

Traditional medicine uses the leaves, bark, roots, fruits, and seeds to cure a variety of ailments, despite the lack of clinical evidence that these treatments are safe or beneficial. Aegeline is a recognized component of the bael leaf that is used as a dietary supplement to help people lose weight.

Dosing

The right dose of bael is determined by many factors, including the user's age, health, and other circumstances. There is currently insufficient scientific data to define a suitable dosing range for bael. Keep in mind that natural products aren't always safe, and that dosages are crucial (Kovach et al., 2009). Before using, be sure to read the product label and consult your pharmacist, physician, or other healthcare experts.

Baneberry

Synonyms

Actaea *alba*, Actaea pachypoda, Actaea rubra, Actée Blanche, Actée Pachypoda, Baneberry, Cohosh Blanco, Coralberry, Doll's Eye, Snakeberry, White Baneberry.

Medicinal Uses and Extracts

Menstruation stimulation (periods).

Taking care of female issues.

Colds.

Coughs.

Problems with the stomach.

Other circumstances.

Dosing and Side effects.

White cohosh is extremely dangerous. The plant is harmful in all sections. Stomach issues, vomiting, bloody diarrhea, headaches, heart and blood circulation disorders, and delirium are all possible side effects.

Ban Xia

Synonyms

Pinellia is a flowering plant in the Araceae family native to China. It is indigenous to Japan and southern China, and its blossom is known as the "green dragon" owing to its color and shape.

Medicinal Uses and Extracts

In Chinese medicine, Ban Xia is one of the most significant herbs for drying dampness and changing cold phlegm. While Ban Xia is frequently used in formulations to help remove phlegm from the lungs, it is most known for its ability to change cold phlegm situations in the spleen and digestive tract. Chronic insufficiency or hypofunctioning of the digestive tract sometimes results in thin, chilly phlegm. This is generally attributed to food intemperance and overeating in Chinese culture (Karmakar et al., 2010).

The root's spicy and warm character resonates with the lung, spleen, and stomach channels. Ban Xia's functions in the body aid to descend rebellious qi, which can induce stomach distress and vomiting, in addition to drying moisture. Ban Xia is essential for relieving distension and stuffiness in the chest and belly since it efficiently enhances directionality in the digestive system. It can help with Lyme Disease-related brain fog, fluid retention, and the sensation of a lump in the throat that comes with emotional discomfort. It can also be used topically to treat skin ulcers and ulcerations. It is dried, powdered, and then mixed with egg white for this purpose.

Side Effects

When consumed by mouth, *Pinellia ternata* is POSSIBLY UNSAFE for the vast majority of people. It features ephedrine, a stimulant. Ephedrine has the potential to produce serious side effects like heart attack, stroke, or seizures. This product is prohibited in the United States due to these safety concerns (Asnake et al., 2016).

Dosage

Pinellia ternata dosage is determined by a number of factors, including the user's age, health, and other circumstances. There is currently insufficient scientific data to define an appropriate dosing range for *Pinellia ternata*.

C

Catechu

Synonyms

It is also known as cutch, black cutch, cachou, cashoo, terra Japonica, or Japan earth, and also katha in Hindi, kaath in Marathi, khoyer in Assamese and Bengali, and kachu in Malay (hence the Latinized

Medical Uses and Extracts

Catechu is most typically taken by mouth to treat stomach issues such as diarrhea, colon enlargement (colitis), and indigestion. It's also used topically to relieve pain, bleeding, and swelling caused by osteoarthritis (inflammation). However, there is little scientific data to back up any of these claims.

Catechu is used as a flavor ingredient in foods and beverages.

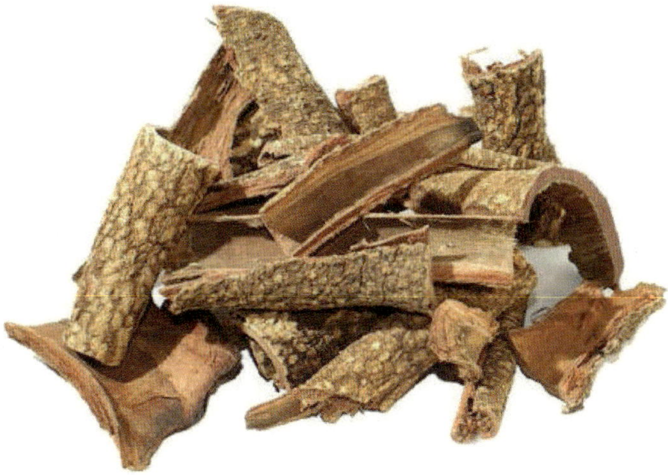

Figure 14: Catechu.

Source: https://www.indiamart.com/proddetail/acacia-catechu-bark-15331284173.html

Dosage

The right dose of catechu is determined by a variety of factors, including the user's age, health, and other circumstances. There is currently insufficient scientific data to define an appropriate dosing range for catechu. Keep in mind that natural products aren't always safe, and that dosages are crucial. Before using, be sure to read the product label and consult your pharmacist, physician, or other healthcare experts.

Side Effects

Catechu is LIKELY SAFE when taken by mouth in the levels found in food. When consumed in medical levels for a short length of time, Catechu is POSSIBLY SAFE. Flavocoxid (Limbrel, Primus Pharmaceuticals), a combination medication including catechu, was found to be safe in research investigations lasting up to 12 weeks (Kayani et al., 2015).. However, some people are concerned that this combined product may cause liver problems. This adverse effect does not appear to be widespread, and it may only be experienced by those who have an allergic reaction to it.

Calendula

Other Names

Calendula (*Calendula officinalis*), sometimes known as pot marigold, is a flowering plant. It is not the same as Tagetes genus decorative marigolds cultivated in vegetable gardens.

Medical Uses

Calendula is a flowering plant native to Asia and southern Europe that has long been utilized in Ayurvedic and Unani therapy. Calendula's compounds may aid in the growth of new tissue in wounds and reduce swelling in the mouth and throat.

Extracts

Calendula flower is used to treat wounds, rashes, infections, inflammation, and a variety of other ailments. However, there isn't much data to back up the usage of calendula for any reason (Kassam et al., 2010).

Side Effects

Calendula flower concoctions are generally safe for most people when consumed by mouth.

Calendula flower solutions are generally safe for most people when applied to the skin.

Dosing

Pot marigold florets are edible. They are often used to add color to salads or added to dishes as a garnish and in place of saffron. The leaves are edible but are often not palatable. They have a history of use as a potherb and in salads. The plant is also used to make tea

Camellia Sinensis

Synonyms

Camellia sinensis is a flowering plant in the Theaceae family with evergreen shrubs or small trees whose leaves and leaf buds are used to make tea. "Tea plant," "tea shrub," and "tea tree" are all common names for this species

Medicinal Uses

Their catechin and L-theanine contents may help lower risk factors for heart disease, including:

- Blood cholesterol and triglycerides
- Arterial plaque.
- Blood pressure.

The extract's cancer-fighting properties are mostly attributed to its catechins, particularly EGCG.

Epigallocatechin gallate (EGCG) is a rare plant chemical that has received a lot of attention due to its possible health benefits.

Among other benefits include

- Effects on stress and anxiety
- Effects on weight loss

Side Effects

Although *Camellia sinensis* leaf extract is generally safe and well-tolerated, there have been rare reports of unfavorable side effects.

Its caffeine level, for example, may cause jitteriness and headaches.

Dosage

Green tea contains at least 180 mg of catechins and 60 mg of theanine in a daily dose of 3 to 5 cups (720 to 1,200 mL). Because of the risk of hepatotoxicity from high doses of epigallocatechin gallate, green tea extract ought not to be taken on an empty stomach (EGCG) (Jiang et al., 2019).

Extracts

The oil derived from the leaves of the *Camellia sinensis* plant is known as Camellia sinensis leaf extract. It's high in bioactive chemicals like catechins, L-theanine, and caffeine, and research has connected it to a variety of health benefits. It may also help you lose weight by boosting your immune system, and reducing stress and anxiety.

Cannabis Sativa

Synonyms

Bud, ganja, grass, hashish, hemp, Indian hemp, marijuana, pot, reefer, weed.

Medical Uses

Cannabis sativa, popularly known as hemp, is a plant that belongs to the Cannabinaceae family. THC (delta-9 tetrahydrocannabinol) is a chemical component found in cannabis that is thought to be responsible for the majority of the psychoactive effects of cannabis, including the "high" that is felt when it is eaten. Cannabis, on the other hand, does not only contain psychoactive components (Jima & Megersa, 2018).

Overall, medical cannabis research has been of poor quality, and it is unclear whether it is an effective treatment for any ailment or whether the risks exceed the benefits. It does not appear to help with chronic pain or muscular spasms, according to the data. Its usage for lowering nausea during chemotherapy, improving appetite in HIV/AIDS, promoting sleep, and improving tics in Tourette syndrome is based on low-quality evidence. Cannabinoids have also been prescribed for anorexia, arthritis, glaucoma, and migraines when traditional treatments have failed.

Side Effects

There isn't enough evidence to draw firm conclusions on the safety of medical marijuana. Tiredness, dizziness, increased appetite, as well as cardiovascular and psychotropic effects, are common side effects of medical cannabis use. Short-term memory impairment, poor motor coordination, changed judgment, and, at high doses, paranoia or psychosis are all possible side effects. Over days or weeks, tolerance to these effects develops.

Dosage

- You may easily establish the dosage using pre-rolled joints if you enjoy the classics and the aesthetics of smoking dry cannabis flowers. For novices, a dose of 0.25 mg to 0.5 g is recommended.
- Edible marijuana has a significantly greater impact on the body than smoking marijuana. Yes, all of these adorable, delicious pies have the potential to transport a reckless consumer into deep space, where he or she is unlikely to be happy. Only people

who have been using marijuana for a long time and are regularly increasing their dosage are suited for such research.

- This is yet another potent cannabis-derived substance. It is even more powerful than food in its purest form.

Extracts

While cannabinoid extracts generated from hemp are well-known for their psychotropic and therapeutic qualities, other functional properties such as the nutritional value of hemp seed oil and the antimicrobial capabilities of hemp extract are frequently disregarded (Helmstädter & Staiger, 2014). Both the cannabis and hemp sectors are interested in isolating bioactive chemicals from hemp and converting them into products that can be used for many applications, from nutritional supplements to antimicrobials, as well as breakthroughs in the administration of therapeutic bioactive.

Carica Papaya

Synonyms

Banane de Prairie, *Caricae Papayae Folium*, *Carica papaya*, *Carica peltata, Carica posoposa*

Medicinal Use

Papaya is used as a sedative and diuretic, as well as to prevent and cure gastrointestinal tract diseases and intestinal parasite infections. Nerve discomfort (neuralgia) and elephantoid growths are also treated with it. Elephantoid growths are swollen, huge portions of the body that are signs of a rare lymphatic system illness caused by parasitic worms.

Papaya includes papain, a substance that is widely used to tenderize meat.

There isn't enough evidence to rate effectiveness in the case of...

- Cancer. Consumption of papaya has been associated to a lower incidence of gallbladder and colorectal malignancies, according to research.
- Diabetes. According to preliminary studies, taking fermented papaya everyday for two months can lower blood sugar levels in diabetics.

- Infection with the human papillomavirus (HPV). Consumption of papaya has been related to a lower incidence of HPV infection, according to research.
- Problems with the stomach and intestines.
- Infections caused by parasites.
- Other circumstances.

Dosage

The appropriate dose of papaya for use as treatment depends on several factors such as the user's age, health, and several other conditions. At this time there is not enough scientific information to determine an appropriate range of doses for papaya (Gupta et al., 2018).

Side Effects

When papaya is consumed in large quantities or applied to the skin as papaya latex, it is POSSIBLY UNSAFE. Large amounts of papaya consumed by mouth may cause injury to the esophagus. In some people, applying papaya latex to the skin might cause significant irritation and allergic reactions.

Cassia (Senna) Alata

Synonyms

Senna alata is a medicinal tree as well as an ornamental blooming plant belonging to the Caesalpinioideae subfamily. Emperor's candlesticks, candle bush, candelabra bush, Christmas candles, empress candle plant, ringworm shrub, or candle tree are all names for the same plant. It was frequently separated into its genus, Herpetica, as a peculiar species of Senna.

Medicinal Uses

Senna alata (also known as *Cassia alata*) is commonly referred to as the ringworm bush due to its potent fungicidal capabilities in the treatment of ringworm and other fungal skin ailments. To make "green cotton wool," the leaves are mashed in a mortar. This is combined with the same amount of vegetable oil and applied to the affected area twice or three times daily. Every day, a new preparation is made. Yellow chrysophanic acid is one of the active components (Griffiths et al., 2002).

Its laxative action is well documented, thanks to the anthraquinone concentration.

In the Philippines, Senna alata is known as akapulko and is utilized as a decorative and therapeutic plant due to its laxative, purgative as well as antifungal properties.

Side Effects

When eaten by mouth for a short period of time, senna is LIKELY SAFE for most adults and children over the age of two. Senna is a nonprescription drug that has been approved by the FDA. Senna can cause stomach pain, cramps, and diarrhea, among other things.

When taken by mouth for a long time or in excessive quantities, senna is POSSIBLY UNSAFE. Senna should not be used for longer than two weeks. Longer use may cause the bowels to cease working normally, resulting in a need for laxatives. Long-term use can alter the amount or balance of certain substances in the blood (electrolytes), which can result in cardiac problems, muscle weakness, liver damage, and other negative consequences.

Dosage

- Constipation: The average dose for general constipation is 17.2 mg per day. Take no more than 34.4 mg twice a day. 17 mg per day has been utilized in older persons. 28 mg in two separate doses has been used to treat constipation caused by pregnancy.
- Senna doses of 75 mg of sennosides administered the day before the colonoscopy, or 120-150 mg taken once or twice the day before the colonoscopy, have been utilized for bowel preparation.
- The normal dose for children aged 12 and up is 2 tablets per day, each containing 8.6 mg sennosides. 4 pills (34.4 mg sennosides) twice daily is the maximum dose (Fitzgerald et al., 2020). The normal dose for children aged 6 to 11 years is one tablet (8.6 mg sennosides) per day. 2 tablets (17.2 mg sennosides) twice daily is the maximum dose. The normal dose for children aged 2 to 5 years is 1/2 tablet (4.3 mg sennosides) per day. 1 pill (8.6 mg sennosides) twice daily is the maximum dose.

Extracts

Senna is a type of herb. The plant's leaves and fruit are used to produce medication.

Senna is an OTC laxative that has been approved by the FDA. Senna can be purchased without a prescription. It's used to treat constipation and to cleanse the bowels before diagnostic procedures like colonoscopies. Sennosides are a group of compounds found in senna. Sennosides irritate the intestinal lining, resulting in a laxative action.

Cocos Nucifera

Synonym

The coconut tree (*Cocos nucifera*) is the only extant species of the genus Cocos, which belongs to the palm tree family Arecaceae. The name "coconut" (or the ancient "cocoanut") can refer to the entire coconut palm, the seed, or the fruit, which is a drupe, not a nut, according to botanical classification.

Medical Uses

Coconut palm is said to have medicinal properties. The roots have antipyretic and diuretic properties. Young coconut milk has diuretic, laxative, antidiarrheic, and anti-poisoning properties. The oil is used to heal sick skin and teeth, as well as to make embrocations when blended with other treatments.

Extracts

Coconut milk, which should not be confused with coconut water, is made by pressing shredded coconut meat with boiling water to extract the coconut oil, proteins, and aromatic compounds. It's used to make a variety of foods. Coconut milk has a fat content ranging from 5% to 20%, while coconut cream has a fat content ranging from 20% to 50% (Eddouks et al., 2017). The majority of it (89%) is saturated fat, with lauric acid as the main fatty acid. Coconut milk beverages can be made by diluting coconut milk. These are substantially lower in fat and can be used as a milk substitute.

After centrifugation, separation, and spray drying, coconut milk powder, a protein-rich powder, can be extracted from coconut milk.

Side Effects and dosage

When consumed in food quantities, coconut is LIKELY SAFE. When used as medicine, coconut is POSSIBLY SAFE. Coconuts may produce an allergic reaction in certain adults and children. Rashes on the skin and respiratory problems are common symptoms.

The right amount of coconut depends on a variety of factors, including the user's age, health, and other circumstances. There is currently insufficient scientific evidence to define an appropriate dosing range for coconut.

Cinnamomum Camphora

Synonyms

Camphor tree, camphorwood, and camphor laurel are all names for the same type of evergreen tree Cinnamomum camphora.

Medical Uses and Extractions

- Coughs. Camphor in concentrations less than 11 percent has been cleared by the FDA for use as a chest massage.
- Acute discomfort. Camphor has been licensed by the FDA for use on the skin as a pain reliever in doses ranging from 3% to 11%. Many rub-on lotions contain it to help with discomfort from cold sores, insect stings and bites, mild burns, and hemorrhoids (Dhama et al., 2018).
- Itching. Camphor is FDA-approved for application on the skin in quantities ranging from 3% to 11% to relieve itching and irritation.
- Bite of an insect. According to a preliminary study, combining camphor with menthol and eucalyptus oil may help reduce the size of mosquito bites.
- Acne.

Side Effects

Camphor is extremely dangerous when taken by mouth. Camphor consumption can result in serious negative effects, including death.

When administered to the skin, camphor is POSSIBLY SAFE for most adults when used in low amounts in a cream or lotion. Camphor can cause minor adverse effects like skin irritation and redness. Do not use camphor products that are undiluted or contain more than 11 percent camphor (Devkota & Watanabe, 2020). These can be both unpleasant and dangerous. When applied to damaged or injured skin, camphor-containing products are LIKELY UNSAFE. Camphor is quickly absorbed via broken skin and can accumulate in the body in dangerous quantities.

Dosing

When inhaled as the vapor from rubs applied to the chest, camphor, it is POSSIBLY SAFE for most individuals.

To treat a cough, apply a thin coating of 4.7 percent to 5.3 percent camphor ointment to the throat and chest. A warm, dry towel can be used to cover the region or it can be left uncovered.

For itching, apply a 3 percent to 11 percent ointment three to four times each day.

A 3 percent to 11 percent ointment is commonly used three to four times daily for acute discomfort.

Extracts

Camphor oil is a steam-distilled oil that is derived from the wood of camphor trees. It can be applied topically to treat pain, itch, and irritation. Camphor is also used to treat inflammatory disorders and chest congestion. It has a strong odor and flavor, and it absorbs quickly through the skin.

Citrus Acida

Synonyms

The Key lime (*Citrus aurantifolia*) is a tropical Southeast Asian citrus hybrid (*Citrus hystrix, Citrus medica*). It has a spherical fruit with a diameter of 25–50 mm (1–2 in). The Key lime is typically selected while still green, but when fully mature, it turns yellow.

Medical Uses and Side Effects

- Blood pressure. Women in their forties and fifties who consume more lemon juice or eat more lemons appear to have lower systolic blood pressure (the top number). They do not, however, appear to have reduced diastolic blood pressure (the bottom number).

- Heart attack. Early study suggests that smelling the scent of lemon while in the hospital after a heart attack can help the heart heal by lowering blood pressure and anxiety (Douwes et al., 2008).
- Nausea and Anxiety. A small study found that putting lemon essential oil on a cotton ball and inhaling the aroma can help pregnant women feel less nauseated.
- Radiation therapy has caused problems with saliva production.
- Swelling is reducing.
- Infection of the airways.
- Hay fever.
- Kidney stones.
- Increasing urine.
- Meniere disease
- Obesity.
- A disease caused by vitamin C deficiency (scurvy).

Dosage

The right amount of lemon to use as a treatment depends on a variety of factors, including the user's age, health, and other circumstances. There is currently insufficient scientific evidence to define a suitable dosing range for lemon. Keep in mind that natural products aren't always safe and that dosages are crucial. Before using, be sure to read the product label and consult your pharmacist, physician, or other healthcare experts.

Extracts

These findings reveal that citrus methanolic extract has significant antioxidant activity that is nearly equal to synthetic antioxidants (BHA and BHA). As a result, using citrus peel extract as a natural antioxidant to prevent rancidity in oils and fats is recommended.

Chilli

Synonym

The chili pepper (also chile, chile pepper, chilli pepper, or chilli) is the berry-fruit of plants belonging to the nightshade family, Solanaceae, from the genus Capsicum. Chili peppers are commonly used as a spice to lend a pungent 'heat' to meals in various cultures (Do et al., 2019). When ingested or administered topically, capsaicin and related compounds known as capsaicinoids are the ingredients that give chili peppers their heat.

Medicinal Uses and Extracts

Chili pepper possesses anti-cancer and anti-inflammatory characteristics, as well as anti-inflammatory and anti-inflammatory properties for rheumatism, stiff joints, bronchitis and chest colds with cough and headache, arthritis, cardiac arrhythmias, and stomachic properties. Standard capsaicin, pharmaceutically produced gels, lotions, and plasters, essential oils distilled from pods, a powder made by crashing pods, and extracts made by soaking pods in water or ethanol are all options for treating diseases using chili.

Dosage

Capsicum extracts, chili, fermented red pepper, red pepper powder, lozenges, nectar, and capsaicin have all been consumed by adults. Capsicum has been used on the skin of both adults and children in creams, plasters, and gels. Nasal sprays and solutions also include it. Consult a healthcare professional to determine the optimal product and dosage for a specific problem.

Side Effects

Capsicum is likely safe when consumed in amounts found in food when taken by mouth. Short-term usage of capsaicin, the primary ingredient in capsicum, may be harmless. Stomach irritation, sweating, and a runny nose are all possible side effects. Capsicum may be harmful if used in big dosages or over an extended period of time (Cummings & Olsen, 2011).

When applied to the skin, capsicum extract-containing lotions and creams are considered to be safe for most adults. The FDA has approved capsaicin as an over-the-counter drug. Skin inflammation and itching are possible side effects. Capsicum has a strong irritant effect on the eyes, nose, and throat. Capsicum should not be used on sensitive skin or near the eyes.

Chick Pea

Synonyms

Chickpeas (*Cicer arietinum*) are an annual legume in the Fabaceae family, subfamily Faboideae. Gram or Bengal gram, garbanzo or garbanzo bean, or

Egyptian pea are some of the names given to it. Chickpea seeds are a good source of protein. It is one of the first legumes to be farmed, with 9500-year-old remains discovered in the Middle East.

Medical Use and Extracts

Chickpeas help control blood sugar.

They help with digestion.

They can help lower cholesterol.

They may lower your cancer risk.

They give you stronger bones.

They could boost your mental health.

Dosage and Side Effects

Saponins come with several drawbacks. All legumes include saponins, which are natural chemical substances. Aquafaba is the foam that forms in the liquid made from canned chickpeas (Carvalho & Morales, 2010.

Saponin-containing plants can cause stomach distress and diarrhea in some animals. According to studies, they can harm female mice's reproductive systems, especially their ovaries. They're also poisonous to fish and cold-blooded animals like snails.

Dosage for chickpea is dependent on size age and various factors as it is a food product ordinarily consumed

D

Daikon/Radish

Raphanus sativus var. longipinnatus, also known as daikon (Japanese for "big root") or mooli, is a mild-flavored winter radish with fast-growing leaves and a lengthy, white, napiform root. Daikon is a vegetable native to continental East Asia that is now grown and eaten all through the region, as well as in South Asia, and is now available worldwide. Daikon is grown for its ability to break up compacted soils and recover nutrients in some areas, although it is not harvested.

Daikon is a good source of potassium, magnesium, calcium, and copper, among other nutrients. Nonetheless, it has the highest levels of vitamin C and folate. Vitamin C is a water-soluble vitamin that is necessary for a variety of body activities, including immune system function and tissue growth and repair. It is also a potent antioxidant that protects your body's cells from oxidative harm (Bonini et al., 2018). When consumed by mouth, radish is LIKELY SAFE for most people when consumed in moderation. Large amounts of radish might irritate the stomach and intestines. Some people may be allergic to radish, but this is uncommon. The right amount of radish to take depends on several factors, including the user's age, health, and other circumstances.

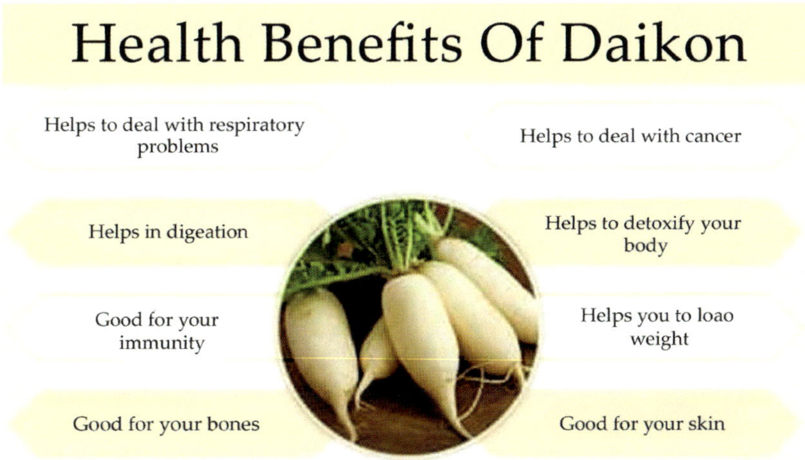

Figure 15: Daikon juice, as well as the roots and leaves of the daikon plant, contain anti-inflammatory chemicals that aid to reduce inflammation throughout the body. This has a lot of beneficial impacts on your body, including maintaining heart health, lowering the risk of developing arthritis, treating gout, and reducing the discomfort and suffering associated with accidents and muscle strains.

Source: https://www.lybrate.com/topic/benefits-of-daikon-and-its-side-effects

Daikon Seed Extract is an emollient oil made from daikon radish seeds that is very light in color and odorless. This unique and outstanding emollient oil has been generated from these seeds specifically for the beauty industry and has long been farmed in Oregon to supply Asian countries with radish seeds for sprouts and veggies. Daikon Seed Extract, which is made up of an unusual blend of C18, C20, and C22 fatty acids, provides formulators with unique chemistry (Braun, 2005). It provides a natural alternative to silicones and emollient esters for imparting slide, good absorption, and a non-oily feel to the skin and hair care products, with a sensory profile more closely linked with these chemicals.

Damiana

Also called Rosemary, Oreganillo, Old Woman's Broom, Mizibcoc, Mexican Damiana, Houx Mexicain, Herba de la Pastora, Damiane, Damiana leaf, Damiana herb, and Damiana Aphrodisiaca, Damiana is a shrub native to Mexico, Central America, and the Caribbean. Medicine is made from the leaf and stem. It was traditionally used to boost sexual desire (as an aphrodisiac). Damiana is used to cure headaches, bedwetting, melancholy, anxious stomach, and constipation, as well as to prevent and treat sexual issues, increase mental and physical stamina, and act as an aphrodisiac.

Damiana dosage is determined by several factors, including the user's age, health, and other circumstances. There is currently insufficient scientific data to define an appropriate dosing range for damiana. Keep in mind that natural products are not always safe and that dosages are crucial. When consumed by mouth at proportions usually found in meals, Damiana is LIKELY SAFE. When taken by mouth in therapeutic proportions, Damiana is POSSIBLY SAFE, although there have been major negative effects. After consuming 200 grams of damiana extract, people have reported convulsions and other symptoms comparable to rabies or strychnine poisoning (Bhattacharyya, 2015).

Although its major historical use in cultures has been to increase sexual desire, it has also been prized for its relaxing and calming properties. Damiana leaf has also been used as a tonic, a tasty beverage, and to improve the daily functioning of the digestive system. Damiana leaves are commonly used in tea blends, herbal smoking mixes, infused liquors, and macerated as damiana extract.

Dandelion

The common dandelion, *Taraxacum officinale*, is commonly thought of as an annoyance and difficult-to-control weed, but it's also a wonder weed with numerous health advantages ranging from improved digestion to psoriasis control. For centuries, dandelion has been utilized as a skin toner, stomach remedy, laxative, and treatment for PMS, warts, arthritis, and kidney stones. It was first described as medicine by Arabian doctors in the 10th and 11th centuries, and subsequently by Welsh healers in the 1200s. Due to the obvious tooth-like serrations on the plant's leaves, which funnel water to the root, it gets its name from the French dent de lion, which means "teeth of the lion."

Dandelion coffee is made by roasting the root itself, which is caffeine-free. Teas, Tinctures (liquid formed from a plant), Decoctions (infusions), Poultices (a paste created from plants), and Capsules can all be made from the dried or fresh root when used for medicinal (Blancke, 2016). Dandelion root has traditionally been used to heal stomach and liver problems in traditional Chinese and Native American medicine. Many health problems, including cancer, diabetes, gastrointestinal diseases, heartburn, high cholesterol, acne, and eczema, are believed to be treatable by herbalists today.

When consumed in moderation, dandelion root is generally regarded safe and well-tolerated by adults. Heartburn, diarrhea, upset stomach, and irritated skin are some of the possible adverse effects. Dandelion root allergy sufferers may develop a rash, eye irritation, and other allergy symptoms. Dandelion also includes iodine and latex, so it's best to avoid it if you're allergic to either. Because there isn't enough data on the long-term safety of dandelion treatments, pregnant women, nursing mothers, and children should avoid them. It is quite likely that eating too much dandelion will lower women's fertility and men's testosterone levels. This could be caused by phytoestrogen, a plant chemical that acts like estrogen.

Figure 16: The leaves are used to aid digestion and boost appetite. The blossom of the dandelion is high in antioxidants. Dandelion may also aid in the improvement of one's immune system. Dandelion roots and leaves are used by herbalists to detoxify the liver and gallbladder, while dandelion leaves are used to support kidney function.

Source: https://www.google.com/url?sa=i&url=https%3A%2F%2Fwww.medindia.net%2Fpatients%2Flifestyleandwellness%2Fhealth-benefits-of-dandelion-plant.htm&psig=AOvVaw3ZpkGuKRXwBvz9ngSqOD2e&ust=1633777252215000&source=images&cd=vfe&ved=0CAwQjhxqFwoTCIilxdnUuvMCFQAAAAAdAAAAABAD

In the United States, there are no regulations for the proper use of dandelion root. In Europe, nevertheless, the European Commission and the British Herbal Pharmacopoeia recommend the appropriate dosages for adults: 2 to 8 grams of fresh dandelion root each day; 3 to 4 granules of powdered dandelion root per 150 milliliters of warm water Infuse 1 spoonful of minced root in 150 mL of boiling water for 20 minutes to make a dandelion tea infusion. 1 to 2 tablespoons of fresh root extract every day: and 0.75-to-1.0-gram dried dandelion extract per day (Bussmann & Sharon, 2006). Dandelion root supplements can also be found in pharmacies and vitamin supplement stores. In these types of stores, you can also obtain tinctures, teas, extracts, ointments, powders, and dried organic roots. As a general guideline, never exceed the dosage prescribed by the manufacturer. Stop taking the medication and contact your doctor if you notice any negative effects.

Devil's Claw

Also called Devil's claw root, Grapple plant, Garra del Diablo, Racine de Windhoek, Wood spider, Harpagophytum, and *Uncaria procumbens*, Devil's claw is a medicinal plant. Harpagophytum is a Greek word that means "hook plant." This plant, which is native to Africa, derives its name from the look of its fruit, which is coated in hooks intended to hook onto animals to distribute the seeds. The roots and tubers of the plant are used to manufacture medication.

Devil's claw is used to treat atherosclerosis (hardening of the arteries), gout, arthritis, muscular pain (myalgia), backache, tendonitis, chest discomfort, GI upset or heartburn, fever, and severe headaches. It's also used to treat issues like infertility, menstruation irregularities, allergic reactions, appetite loss, and bladder and kidney disorders. For injuries and various skin disorders, some individuals apply devil's claw to the skin (Agbogidi & Azagbaekwe, 2013). When taken by mouth in suitable doses for up to a year, devil's claw is POSSIBLY SAFE for most individuals. Diarrhea is the most prevalent side effect. In one trial, almost 8% of those who took part in it got diarrhea. Nausea, vomiting, abdominal discomfort, migraines, ringing in the ears, lack of appetite, and loss of taste are all possible adverse effects. It can also cause adverse skin reactions, menstruation issues, and blood pressure fluctuations.

A specialized powdered devil's claw root product (Harpadol, Arkopharm) dosed at 2.6 grams per day for osteoarthritis. This dosage contains a total of

57 mg of harpagoside and 87 mg of total iridoid glycosides, both of which are active substances. Another devil's claw extract (Doloteffin, Ardeypharm) was utilized at a dose of 2400 mg/day. For back discomfort, a particular devil's claw extract (Doloteffin, Ardeypharm) with 50-100 mg of the active component harpagoside per day is recommended.

The level of active chemicals in devil's claw preparations varies depending on the amount of harpagoside present. The daily dose of Doloteffin (extraction solvent water) contains twice as much harpagoside as preparations extracted with 60% ethanol. Only preparations that are beneficial for uncomfortable lower back or arthritic pain are a viable alternative to synthetic analgesics in the treatment of chronic pain. Extracts containing at least 50 mg harpagoside in the daily dose should be suggested for the treatment of pain, according to the data.

Dill Dong quai / Angelica

Also called Don Quai, Kinesisk Kvan, Ligustilides, Radix Angelicae Gigantis, Angelica China, *Angelica sinensis*, *Angelica polymorpha var. sinensis*, Angelicae Gigantis Radix, and Chinese Angelica, Dill Dong Quai is plant, and its root is used to make medicine. Dong quai is often used by mouth for menopausal problems, menstrual cycle diseases like migraines, and a variety of other ailments, however there is little scientific data to back up these claims. Animal studies have indicated that dong quai root affects estrogen and other hormones. It's unclear whether these similar consequences occur in humans.

Various dong quai-containing combo products appear to alleviate menopausal symptoms. In menopausal women, taking a specific product containing dong quai and chamomile (Climex) appears to lessen hot flushes (Athar & Nasir, 2005). In pre-and postmenopausal women, using a specific product containing American ginseng, red clover, milk thistle, black cohosh, dong quai, and vitex agnus-castus (Phyto-Female complex) appears to minimize hot flushes and nocturnal sweats while also improving sleep quality. When administered directly to the skin of the penis in conjunction with other herbs, it prevents premature ejaculation. According to preliminary studies, an injection of a product combining dong quai and other herbs may alleviate chest pain and enhance heart function in persons with heart disease.

Figure 17: Dong quai advantages have long been recognized throughout Asia, where the root is also known as Radix angelicae sinensis and has been used as an everyday tonic and herbal medicine for generations.

Source: https://www.herbazest.com/herbs/dong-quai

Dong quai is POSSIBLY SAFE for adults when taken by mouth for up to 6 months. It's usually used at a 100-150 mg daily dose in combination with other substances. It can make your skin hypersensitive to the sun. This could make you more susceptible to sunburn and skin cancer. Outside, use sunblock, especially if you have fair skin. It is POSSIBLY UNSAFE to take dong quai in larger doses for more than 6 months. Dong quai contains cancer-causing substances. When used on the skin, there is insufficient credible evidence to determine whether dong quai is safe or what the potential adverse effects are. The correct dosage of dong quai is determined by several factors, including the user's age, health, and other circumstances. There is currently insufficient scientific evidence to define an optimal dosing range for dong quai. For its wellness-promoting properties, dong quai root powder is used both alone and in combination herbal formulations, and it has even been referred to as "female ginseng" in Traditional Chinese Medicine. Powdered dong quai can be tinctured and encapsulated.

Desmodium Triflorum

Desmodium triflorum is also known as Three-flower beggarweed and Creeping tick trefoil. The United States, China, Malaysia, Australia, and Southeast Asia are among the tropical and subtropical locations where this plant can be found (including India, and Bangladesh). *Desmodium triflorum* is believed to have originated in Asia's tropical regions. *Desmodium triflorum* can be found in a variety of habitats, including waste areas, disturbed and farmed regions, woodland borders, and pastureland.

This plant is continuously being studied because it has a wide range of medical qualities. It may be capable of curing a variety of ailments, including tuberculosis and cancer. However, when additional study is completed, we can anticipate learning more about *Desmodium triflorum* (Zhou et al., 2014). This plant was used by some ancient people as an herbal cure for diarrhea, painful urination, and anti-inflammatory effects. *Desmodium triflorum* is a diuretic used to alleviate aches and skin rashes in traditional medicine. It also aids in the improvement of appetite. This plant was also used to treat syphilis and gonorrhea. It improves the digestive system from the inside out. *Desmodium triflorum* contains diuretic characteristics, which aid in the removal of toxins from the body via urine. Desmodium comes in a variety of forms, some of which can be used to treat cancer.

Figure 18: *Desmodium triflorum* is a well-known and widely used medicinal plant in Sri Lanka. Its antioxidant, anti-inflammatory, anti-convulsant, and analgesic properties have been proven in recent pharmacological research. Alkaloids abound in this plant.

Source: http://medplants.blogspot.com/2015/10/desmodium-triflorum-hamsa-padi.html

Desmodium triflorum's toxicity has yet to be thoroughly investigated. There is no evidence to support any of these traditional uses. Toxins found in this plant can harm the liver, kidneys, and even the lungs. The toxic elements of this plant could hurt the skin if anyone receives a wound or cut while working with it. *Desmodium triflorum*, like any other natural plant, can cause allergic reactions in the body. Several toxic plants appear like these in nature. Any plant that resembles Desmodium triflorum could be another dangerous plant.

Decoction

A decoction is a technique of extracting herbal or plant chemicals by boiling the material, which can include bark, roots, stems, and rhizomes, to dissolve the chemicals. A decoction is crushing the plant material to maximize solubility and then boiling it in water to extract oils, volatile organic compounds, and other chemical components. Tisanes, tinctures, and other similar solutions can be made with decoction.

Because of the temperature and/or preparation differences, decoctions and infusions may produce liquids with different chemical properties. Decoctions may contain more oil-soluble chemicals than infusions. Though the word is often only used to describe boiled plant extracts, mainly for medical or scientific purposes, it can also be applied to meats and vegetables to make bouillon or stock (Yi et al., 2016). Decoctions are used in herbalism to remove liquids from hard plant parts like roots and bark. The plant material is commonly cooked for 1–2 hours in 1–5 liters of water to achieve this. After that, it is indeed strained. This procedure is also used in Ayurveda to make Kashayam herbal remedies. Decoction includes boiling the same quantity of herb and water as an extraction (one teaspoon per cup) for five to ten minutes for teas. Decoction mashing is a technique of brewing in which a portion of the mash is transferred to a different vessel, boiled for a period, and then reintroduced to the main mash, increasing the mash temperature to the next temperature level.

Dioscorea Alata

Dioscorea alata is a yam species that goes by various names, including purple yam, ube, and larger yam (a tuber). The tubers are normally a vibrant violet-purple to vibrant lavender color (thus the common name), but others are cream or plain white in appearance. If consumed raw, it can be poisonous. It's a laxative and vermifuge used to treat fever, gonorrhea, leprosy, tumors, and irritating hemorrhoids in medicine. To prevent or delay a possible miscarriage, the tuber is grated, mixed with brown strong vinegar, then smeared onto paper and placed on the small of a woman's back.

There is no scientific proof that *D. alata* has any therapeutic properties, despite its availability as a dietary supplement and use in folk medicine. Supplementing with *D. alata* may cause side effects in women who are taking estrogen, anticoagulant drugs, or who are pregnant or breastfeeding. The use of *D. alata* supplements may cause allergic reactions in some people. Oxalates (486–781 mg/100 g dry matter) are found in high concentrations in *D. alata*, which have been linked to anti-nutritional effects and kidney stone formation (Van Jaarsveld, 2013). Dioscorea species with edible leaves have opposite leaves, whereas those with poisonous leaves have alternate leaves. This species' uncooked tuber is poisonous and is known to cause narcosis. Saponin is present, and boiling the tubers makes them safe to consume.

Dolichos Catjung

Dolichos catjang, also known as *Vigna unguiculata cylindrica*, is an annual plant that grows in a variety of forms, from erect and bushy to prostrate and scrambling. It produces stems that range in length from 15 to 80 cm. In warm temperate and tropical zones, the plant is frequently cultivated for its edible seed. In the Tropics, it is an abundant cropper that can provide a crop all year, especially if new sowings are performed every few months. It is commonly referred to as cowpeas.

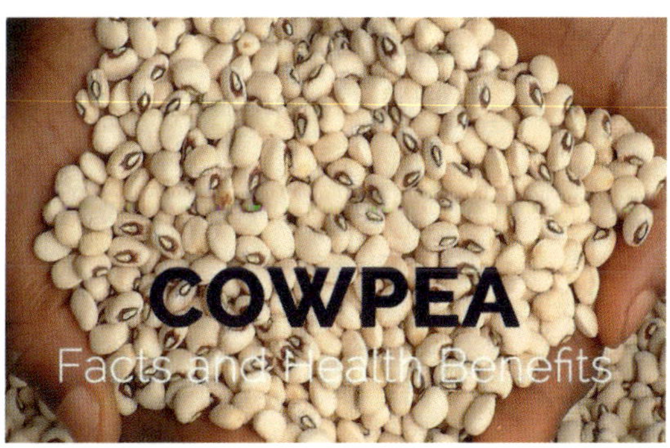

Figure 19: Cowpeas are high in mineral (Iron), which helps to eliminate anemia. Iron aids in protein metabolism, which is necessary for the formation of RBCs and hemoglobin, as well as for preventing anemia. Anemia is caused by a deficiency in hemoglobin and red blood cells.

Source: https://www.healthbenefitstimes.com/cowpeas/

The seed has a diuretic effect. It's a stomach-strengthening supplement. When boiled and consumed as a meal, it is thought to kill stomach worms. Broken bones are treated with a poultice made from crushed leaves. Cowpea seeds are an excellent source of protein, calories, minerals, and vitamins. In regions where cowpeas are a major food crop, this supplemented a primarily cereal-based diet. A seed can contain up to 25% protein and has very little fat. Cowpea starch is metabolized at a slower rate than grain starch, which is better for human health. Folic acid, a vital vitamin that helps prevent neural tube abnormalities in unborn newborns, is abundant in the grain (Umadevi et al., 2012).

Anthocyanins are becoming increasingly popular in a variety of disciplines, including food and pharmaceuticals, due to their antioxidative capabilities. The extraction method has a significant impact on the natural color quality. Cowpea seed coat color is an important factor in customer selection (Vigna unguiculata L. Walp).

Deglycyrrhizinated Licorice

The root of the licorice plant yields licorice root, which is one of the world's oldest herbal treatments (*Glycyrrhiza glabra*). Licorice, which is native to Western Asia and Southern Europe, has long been used to treat a variety of diseases as well as flavor candies, beverages, and medicines. Despite its long history, only a few of its applications have been scientifically validated. Licorice may also pose several health hazards.

Licorice has been used medicinally since ancient Egypt when the root was converted into a sweet drink for pharaohs. It has also been used to ease a stomachache, decrease inflammation, and cure upper respiratory issues in traditional Chinese, Middle Eastern, and Greek medicine. Licorice root is now widely used to treat a variety of diseases, including heartburn, gastroesophageal reflux, hot flashes, cough, and bacterial and viral infections. It's commonly available in the form of a pill or a liquid supplement. Licorice tea is also reported to help with sore throats, while topical gels are said to help with skin issues including acne and eczema. Licorice is also used to flavor a variety of dishes and beverages. Interestingly, many licorice candies are scented with anise oil, an essential oil derived from the anise plant (Pimpinella anisum) that has an identical flavor to licorice root.

Glycyrrhizin is the principal active ingredient in licorice root, which contains hundreds of plant chemicals. The root's sweet taste is due to glycyrrhizin, which also has antioxidant, anti-inflammatory, and antibacterial effects. Glycyrrhizin, on the other hand, has been linked to several licorice root's negative consequences. As a result, some items use deglycyrrhizinated licorice (DGL), which is licorice without the glycyrrhizin.

Figure 20: Licorice root is now widely used to treat a variety of diseases, including heartburn, acid reflux, coughs, hot flashes, and viral and bacterial and infections. It's commonly sold as a pill or a liquid supplement.

Source: https://www.botanical-online.com/en/medicinal-plants/licorice-properties

Licorice root has been declared generally regarded as acceptable for use in foods by the Food and Drug Administration (FDA). Currently, the FDA does not assess or validate supplements for authenticity, efficacy, or ingredient labeling reliability. In addition, licorice root pills and teas are typically deemed safe for short-term use. Large doses, on the other hand, may have negative side effects, and people with specific health issues may want to avoid it. Glycyrrhizin buildup in the body can be caused by both long-term use and excessive doses of licorice root extracts. Glycyrrhizin levels that are too high have been linked to an excessive rise in the stress hormone cortisol, which can lead to fluid and electrolyte abnormalities.

Licorice poisoning is a rare occurrence. It can lead to kidney failure, congestive heart failure, or a build-up of fluid in the lungs (pulmonary edema). Individuals with kidney illness, congestive heart failure, high blood pressure, or low potassium levels should avoid all licorice products containing glycyrrhizin. Licorice root extract is available in a variety of formulations as a supplement, comprising pills, topical gels, tinctures, powder, and beverages (Thakur, 2011). The root itself is available in both fresh and dried form. There is no standard dosage suggestion currently. The World Health Organization (WHO) and the European Scientific Committee on Food (SCF) both advise that glycyrrhizin intake be limited to no more than 100 mg each day.

Dihydrotestosterone

DHT (dihydrotestosterone) is a testosterone metabolite. It produces similar effects, but they are far more powerful. DHT is thought to be three to six times stronger than testosterone. Androgens are the classification for both hormones. DHT is produced predominantly in the organs that need it, such as the prostate. Converting testosterone to DHT is mostly accomplished by the enzyme 5-alpha reductase.

Both testosterone and DHT bind to the same receptors, which is interesting. DHT, on the other hand, does so more effectively. Furthermore, the reactions to DHT and testosterone binding are distinct. The majority of DHT in the body is attached to proteins in the blood, such as albumin and sex hormone-binding globulin. This holds for testosterone as well. In the case of DHT, less than 1% of the hormone is freely circulating in the blood. This is slightly less than the difference between free and bound testosterone.

The consequences of dihydrotestosterone on the human body are numerous. It is also a precursor of other steroid hormones, in addition to being a stronger form of testosterone. Hirsutism and female pattern baldness are other possible side effects. Lack of DHT production, on the other hand, might result in sexual differentiation issues such as androgen insensitivity syndrome. Excessive testosterone production can cause both men and women to suffer with high levels of dihydrotestosterone. High amounts in men generate few noticeable changes. Women with excessive dihydrotestosterone levels, on the other hand, may experience abundant body and facial hair. Adult acne and menstruation may also be a problem for them. A dihydrotestosterone inhibitor may be able to aid in the restoration of natural dihydrotestosterone levels.

Women, on the other hand, experience fewer side effects than men when their dihydrotestosterone levels are low. Low amounts of dihydrotestosterone may cause a boy going through puberty to fail to establish appropriate body hair growth or genital development because its function is so strongly linked to male features. If a young boy is exposed to insufficient amounts of dihydrotestosterone while in the womb, he may be born with ambiguous genitalia that resembles female genitalia.

E

Eleusine Coracana

Eleusine coracana is a tufted annual grass that can reach a height of 170cm. Around 3,000 BC, finger millet was first cultivated as a food crop in East Africa. Cultivation has steadily extended from there and the plant is currently farmed, mostly on a modest scale, in warm temperate to tropical nations around the world. The plant is also utilized as a soil stabilizer and has local medical applications. Local markets sell the seed, but there is little or no international trade.

The seed is cooked. The seed can be cooked whole or processed into flour and served as a millet. It's used in cakes, puddings, and porridge, among other things. If the flour is soaked overnight in water, it makes good unleavened bread. It's frequently utilized in the production of fermented foods. For many populations, finger millet is a staple diet, particularly in drier areas of India, Nepal, and Sri Lanka. Rice, corn, and sorghum are all lower in protein, fat, and minerals than this grain. It provides a sustained diet when ingested as food, especially for persons who work hard. The grain can also be malted, with the malted grain flour being used as a nutritious diet for newborns and invalids.

Figure 21: Finger millet not only aids in maintaining good health and warding off ailments, but it also works wonders in preserving skin's youth and vitality by preventing premature aging. It can be used as the most powerful natural anti-aging beverage.

Source: https://www.healthbenefitstimes.com/finger-millet/

For diabetics, finger millet is a particularly nutritious diet. The seed has a diameter of roughly 2mm. It is possible to have a nutritional analysis. Malt from the seed is commonly used to make local beer and other alcoholic and non-alcoholic beverages. Finger millet grain, except for barley, has a higher enzyme activity in malting, making it ideal for brewing. The grain is said to improve in quality with storage and has great storage qualities. Seeds can be stored for up to 50 years without deterioration. In times of famine, they are highly prized as a reserved meal (Stanisavljević et al., 2007).

The seed is cooling, astringent, and tonic. Fever, biliousness, and hepatitis are all treated with it. The plant is said to be diaphoretic, diuretic, and vermifuge, and the leaf juice has been administered to mothers during labor. Leprosy, liver illness, measles, pleurisy, pneumonia, and smallpox are all treated with the plant. Internal leprosy treatment involves drinking the juice of a blend of finger millet leaves and Plumbago zeylanica plants.

There are no known hazards of Eleusine coracana.

Electuary

Electuary is an old term for an herbal medication that is blended with syrup or honey. It comes in a variety of textures and substances, and it can be consumed as a paste or rolled into small pills. Honey was employed for preservation in kitchens and even monarchs' graves in the days before distillation. The phrase ekleikhein, which means "Lick Out," is thought to have originated in Greek.

An electuary is a medicine made from a powder or other component that has been sweetened with honey to make it more appetizing. Electuary, also known as 'latwerge' in German and Swiss cultures, is a thickened juice and honey concoction with a thick, viscous consistency that is used in cooking as a (bread) spread or a sauce element. A teaspoonful of sweet herbal paste eaten as needed is the normal 'dosage,' however, this can differ depending on the electuary's potency. It's delicious stirred into hot beverages (milk, tea, and hot water), straight from the spoon, or even on buttered whole-grain toast. I've even made medicinal mead and aged drinks using electuaries.

Similar paste-medicines are created in India with Ghee (clarified butter) or Ghee and Honey, with the addition of Coconut Milk or animal milk in some cases. Electuaries are frequently complex medicines that can be stored for a long amount of time. They're usually highly concentrated and, because of the Honey or Sugar content, they're usually quite tasty. The Electuary is sometimes left to ferment for months before being stirred a couple of times weekly till it rises. Mithridate and Theriac were examples of this. You can also make your electuary as powerful as you wish. Electuaries are usually created with powdered herbs, however, if you need to add a little tincture, go ahead! The possibilities for creating an electuary are boundless (Seth, 2003).

Elixir

An elixir is a pleasant drink that is ingested orally for medical purposes and is intended to treat one's sickness. An elixir comprises at least one active component that is intended to be consumed orally when used as a medicinal preparation. This combination drug is used to treat symptoms induced by the common cold, flu, allergies, or other respiratory ailments for a short period (such as sinusitis, bronchitis). Watery eyes, itchy eyes/nose/throat, nasal congestion, and sneezing can all be relieved by antihistamines. Decongestants help to reduce the symptoms of a stuffy nose and ear congestion. If you are self-treating with this drug, read the package instructions carefully before using it to make sure it's good for you. Some goods have the same brand names as others, yet they contain different active compounds that have different applications.

You could become ill if you use the incorrect product. If you have any questions regarding your product or how to use it, talk to your pharmacist. Cough-and-cold remedies have not been shown safe or useful in children under the age of six. Unless otherwise directed by a doctor, do not use this product to treat cold symptoms in children under the age of six. Some products (such as long-acting tablets/capsules) are not indicated for children under the age of 12. For further information on how to use your product safely, consult your doctor or pharmacist. This product should not be used to put a child to sleep. These products do neither treat nor decrease the duration of a cold, and they may have dangerous side effects. Follow all dose instructions carefully to reduce the risk of major adverse effects. Other cough-and-cold medications that may contain the same or similar components should not be given.

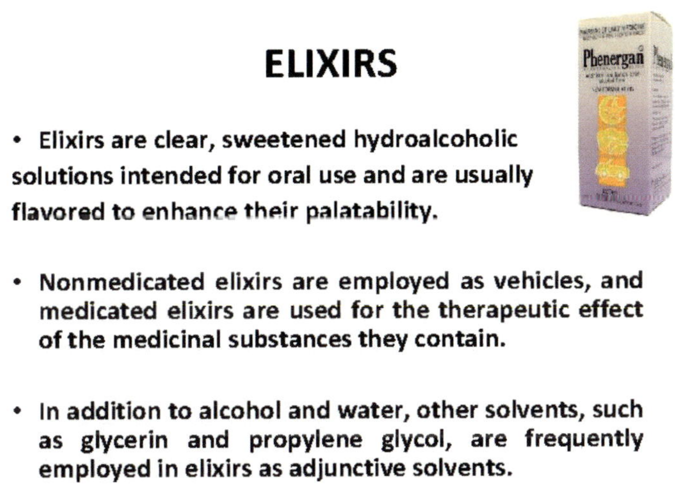

Figure 22: For various administrative routes, solutions can be developed. Orally: Drops, elixirs, and syrups, IV, IM, SC (parenterally) Mouthwashes, gargles, and throat sprays are used in the mouth and throat. Douches, ear drops, enemas, and nasal sprays are used in the body cavities.

Source: https://slidetodoc.com/pharmaceutics-1-liquid-dosage-forms-solutions-1-solutions/

Dosage is determined by the substance you are using, as well as your age, medical condition, and treatment response. Without your doctor's permission, do not raise your dose or take this drug more frequently than prescribed. Improper usage (abuse) of this medicine can have catastrophic consequences (such as seizures, hallucinations, and death). If stomach distress develops, this drug can be given with food. Unless your doctor tells you differently, drink plenty of water. If you're taking this medication in liquid form, use a special measuring device/spoon to accurately measure the dose. If you use a regular spoon, you could not get the right dose. If you're using a suspension, give the bottle a good shake before each dose (Saslis-Lagoudakis et al., 2014). Extended-release pills and capsules should not be crushed or chewed. This can cause the entire medicine to be released at once, increasing the chance of unwanted effects. Also, unless your doctor or pharmacist directs you to, do not divide extended-release tablets unless they have a score line.

Eucalyptus Globulus

Also called Fevertree, Gum Tree, Red Ironbark, White Ironbark, Eucalyptus globulus is a tree that grows only in Australia. Eucalyptus oil is derived from the tree's leaves. Essential oil of eucalyptus is used to treat a range of ailments and conditions, especially nasal congestion, and allergies, and as a tick repellant. Eucalyptus oil, diluted, can also be used on the skin to treat health issues like arthritis and skin ulcers. Eucalyptus oil is also used to help relieve cold symptoms and improve respiratory health. Eucalyptol is extracted from the Eucalyptus globulus tree and is commonly used in mouthwashes and cold treatments. Eucalyptus essential oil is frequently used in diffusers for aromatherapy health benefits.

It is available as an oil. Eucalyptus oil should be diluted and not consumed in its purest form. Considerations for dosage should be given as follows: Arthritis: 5%-20% in an oil-based emulsion and 5%-10% in an alcohol-based composition; and 1 teaspoon in boiling water for steam inhalation; also massage on the chest and around the nose.

Allergic responses are a common eucalyptus adverse effect. If the oil is consumed orally, it can cause severe allergic reactions such as abdominal pain, nausea, and vomiting. Skin discoloration in a blue hue, Breathing problems, coordination issues, etc. Hypotension (low blood pressure), arrhythmias (irregular heartbeats), and multi-organ failure.

Eugenia Caryophyllus

Also called Flores Caryophylli, Flores Caryophyllum, Gewurznelken Nagelein, Girofle, Giroflier, Huile de Clou de Girofle, Kreteks, Lavang, Lavanga, Oil of Clove, Syzygium aromaticum, Tige de Clou de Girofle, Bourgeon Floral de Clou de Girofle, Bouton Floral de Clou de Girofle, Caryophylli Flos, Caryophyllum, Caryophyllus aromaticus, Clavo de Olor, Clous de Girolfe, and Clove Flower, Eugenia cadyophyllus is a type of herb. Oils, dried flower buds, leaves, and stems are used to produce medication.

Clove is used as an expectorant and for stomach distress. Expectorants help to cough up phlegm easier. Clove oil is used to treat diarrhea, hernias, and halitosis. Clove and clove oil are used to treat gas, nausea, and vomiting in the intestines (Nazim et al., 2020). Clove is administered topically to the gums for toothaches, pain relief during dental work, and a consequence of tooth extraction known as "dry socket." It's also used on the skin as a pain reliever and to reduce inflammation in the mouth and throat. Clove is also applied to the skin in combination with other ingredients as part of a multi-ingredient preparation to prevent males from attaining orgasm too early (premature ejaculation).

Clove is used as a flavor in foods and beverages. Clove is used in toothpaste, soaps, cosmetics, fragrances, and cigarette manufacture. Clove cigarettes (also known as kreteks) typically include 60 to 80 percent tobacco and 20 to 40 percent ground clove. One of the compounds in clove, eugenol, works similarly to menthol in reducing the intensity of cigarette smoke.

Clove appears to be SAFE for most people when consumed at meal levels. The safety of ingesting cloves by mouth in larger medical doses is not well understood. When applied to the skin, clove oil or cream comprising clove flower is POSSIBLY SAFE. Clove oil, when used often and repeatedly in the mouth or on the gums, can cause harm to the gum tissue, teeth pulp, epidermis, and oral mucosa. Inhaling clove cigarette smoke or injecting clove oil into veins is VERY LIKELY UNSAFE and can lead to breathing issues and respiratory problems. A dried clove can induce sensitivity and discomfort in the mouth, as well as damage to tooth tissues.

Figure 23: Cloves are a great treatment for stomach disorders. These are beneficial in the treatment of stomach gas, nausea, diarrhea, vomiting, and other digestive system issues. Cloves are also expectorant and effective in cough treatment.

Source: https://www.bimbima.com/ayurveda/medicinal-uses-of-cloves/1261/

To prevent males from having an orgasm too soon (premature ejaculation), apply the following to their bodies: Angelica root, Toad venom, Cinnamon bark, Asiasari root, Torlidis seed, Clove flower, *Panax ginseng* root, Angelica root, *Cistanches deserticola*, and *Zanthoxyl* species, applied to the glans penis one hour before intercourse and wiped off right before intercourse.

Euphorbia Hirta

Also called Snakeweed, Pillbearing Spurge, *Euphorbium Officinarum*, Asthmaplant, *Euphorbia pilulifera*, *Euphorbia hirta, Euphorbia capitulata, Chamaesyce hirta*, and Euforbia, Euphorbe, and *Euphorbia pilulifera, Euphorbia hirta* is a herb. Medicine is made from the portions of the plant that grow above ground. Breathing diseases such as asthma, bronchitis, and chest congestion are treated with euphorbia. Mucus in the mouth and sinuses, throat spasms, hay fever, and tumors are all treated with it. It's been used to induce vomiting in some persons (Tuttolomondo et al., 2014).

Worms, extreme diarrhea (dysentery), gonorrhea, and gastrointestinal issues are all treated with it in India. Some experts have investigated how euphorbia works in animals, but there isn't sufficient data to determine how it works in humans. When taken by oral, it might produce nausea and vomiting, among other things. Do not meet the fresh herb. It might irritate the skin or trigger allergic responses.

If you are pregnant, it is POSSIBLY UNSAFE to take euphorbia by mouth. It has been suggested that it may induce the uterus to contract, which could result in a miscarriage. The stomach and intestines might be irritated by euphorbia. If you have a stomach or intestinal problem, don't use it.

The right dose of euphorbia is determined by several factors, including the user's age, health, and other circumstances. There is currently insufficient scientific data to define a suitable dosing range for euphorbia. Keep in mind that natural products aren't always safe and that dosages are crucial. Before using, be sure to read the product label and consult your pharmacist, physician, or another healthcare expert.

Essential Oil

Aromatherapy, a type of alternative medicine that uses plant extracts to enhance health and well-being, frequently includes essential oils. However, some of the health claims made about these oils are debatable. Compounds derived from plants are known as essential oils. The oils capture the aroma and flavor of the plant or its "essence." Each essential oil has its distinct aroma due to its unique aromatic components. Distillation (through steam and/or water) or mechanical processes, such as cold pressing, are used to produce essential oils. After the aromatic compounds have been removed, they are mixed with a carrier oil to make a finished product. Essential oils obtained through chemical methods are not regarded as authentic essential oils, thus how the oils are generated is vital.

Aromatherapy, in which essential oils are breathed through various techniques, is the most prevalent use of essential oils. Essential oils should never be consumed. Essential oils include molecules that interact with your body in a variety of ways. Some plant compounds are absorbed when applied to the skin. Certain administration techniques, such as applying heat or to diverse parts of the body, are thought to increase absorption. However, study is scarce in this field. Essential oil aromas can excite parts of your limbic system, which is a component of your brain involved in emotions, actions, sense of smell, and long-term memory. Surprisingly, the limbic system plays a significant role in memory formation. This may help to explain why some scents might elicit memories or emotions. The limbic system is also involved in the control of several unconscious physiological activities, including respiration, pulse rate, and blood pressure. As a result, some people believe essential oils have a physical influence on the human body. However, investigations have yet to confirm this. Considering their extensive use, little is understood about essential oils' ability to treat certain illnesses. Oils of superior grade Only employ distillation or cold pressing to extract pure plant components. Avoid oils that have been diluted with synthetic scents, oils, or chemicals. Essential oils, like plants and herbal goods, contain a variety of bioactive components that can be harmful to your health (Teixidor-Toneu et al., 2018). Most essential oils, on the other hand, are deemed harmless when breathed or mixed with a base oil for application on the skin. Keep in mind that people in the vicinity, such as pregnant women, children, and pets, maybe inhaling the odor. Nonetheless, they may cause rashes, asthma attacks, headaches, and allergic reactions, among other things.

While a rash is the most common adverse effect, essential oils can produce more serious responses, and one case of death has been linked to them. Lavender, peppermint, tea tree, and ylang-ylang are the oils most frequently linked to negative reactions. Cinnamon oil, which is high in phenols, can cause skin irritation and should not be used on the skin without first mixing it with a base oil. Meanwhile, citrus essential oils boost the skin's sensitivity to sunshine, increasing the risk of sunburn. Swallowing essential oils is not suggested because it can be hazardous and even fatal in some cases. Few studies have investigated the safety of these oils for expectant or lactating mothers, who are typically advised to stay away from them.

Figure 24: Essential oils are most known for their health advantages, which include calming allergy symptoms, controlling obstructed menstrual cycles, decreasing body odor, avoiding hair loss, digestion, recovering from colds and asthma, and assisting those with breathing problems.

Source: https://www.truerelaxations.com/essential-oils-list-benefits/

If essential oils are blended with a base oil, they are typically regarded as safe to inhale or apply to the skin. They must not be consumed. However, there is little evidence to back up many of their linked health claims, and their effectiveness is frequently inflated. Using essential oils as a supplemental therapy for minor health issues is likely to be safe. If you have a significant health condition or are taking medication, you should talk to your doctor about how to utilize them.

Eicosapentaenoic Acid

Also called Ethyl-EPA, Fish Oil Fatty Acid, Icosapent Ethyl, N-3 Fatty Acid, Omega Fatty Acid, Omega 3, Acide Eicosapentaénoïque, Acide Éthyle-Eicosapentaénoïque, Acide Gras Essentiel, Acide Gras d'Huile de Poisson, Acide Gras Omega, Acide Gras Oméga 3, Acide Gras Polyinsaturé, Acide Gras W3, Acido Eicosapentaenoico, and EPA, Eicosapentaenoic acid is a fatty acid present in cold-water fish meat, such as cod liver, salmon, halibut, tuna, herring, mackerel, whale blubber, and seal blubber.

Eicosapentaenoic acid is a fatty acid that is taken by mouth to treat clogged heart arteries (coronary artery disease), prevent or treat heart attacks, and to lower triglyceride levels in those who have high amounts. It's also used to treat Alzheimer's disease, depression, personality disorder, schizophrenia, and attention deficit hyperactivity disorder, among other mental illnesses (ADHD). It's also used to treat diabetes, asthma, psoriasis, and cystic fibrosis, as well as to prevent vision loss in elderly adults (age-related macular degeneration; AMD).

Eicosapentaenoic acid is also used to treat lung cancer and prostate cancer, as well as to help cancer patients maintain their weight and lessen the negative effects of chemotherapy. Eicosapentaenoic acid is used by women to alleviate the symptoms of menopause, lower blood pressure during high-risk pregnancies, and lessen the danger of an infant's growth being slowed while still in the uterus (Silva et al., 2010).

In fish oil preparations, eicosapentaenoic acid is combined with docosahexaenoic acid (DHA) to treat a variety of conditions, including heart disease prevention and reversal, as well as lupus, lung disease, hay fever, hot flashes, menstrual problems, cancer, asthma, and kidney disease caused by an immune reaction. Migraine headache prevention in teenagers, Raynaud's syndrome, psoriasis, high blood pressure, high cholesterol, Behcet's syndrome, skin infections, rheumatoid arthritis, Crohn's disease, and ulcerative colitis are all conditions treated with this combination.

After surgery, eicosapentaenoic acid is combined with RNA and L-arginine to decrease infections, promote wound healing, and minimize recovery time. It's also used to treat high blood pressure when combined with another fatty acid called gamma-linolenic acid.

Evening Primrose Oil

Evening primrose oil (EPO) is extracted from the seeds of a North American native plant. Bruises, hemorrhoids, stomach issues, and sore throats have all been treated using the plant in the past. Its therapeutic properties could be related to the presence of gamma-linolenic acid (GLA). GLA, or gamma-linolenic acid, is an omega-6 fatty acid present in plant oils. EPO is commonly used as a supplement or as a topical application. Continue reading to understand how EPO may be used to address a variety of current health issues.

Most people are deemed safe to use EPO for a short period of time. The long-term safety of use has yet to be determined. Keep in mind that the Food and Drug Administration does not inspect supplements for quality. When choosing EPO, one should do their homework on the supplement as well as the company that sells it.

Upset stomach, stomach problems, headache, and soft stools are some of the most common EPO side effects. It is likely that taking the smallest amount feasible will help you avoid adverse effects. EPO can cause an allergic reaction in some people. Swelling of the hands and feet, rash, breathing difficulties, and gasping for breath are some of the symptoms of an allergic reaction. If one is on blood thinners, EPO could make one bleed more. If one is using blood pressure drugs or blood thinners, one should not use EPO. The use of topical EPO to help prepare the cervix for delivery is common (Sepúlveda-Cuellar et al., 2021). However, research published by the Mayo Clinic found that taking EPO orally slowed dilatation and was linked to lengthier labor. EPO has not been studied enough to confirm its safety during pregnancy or nursing, thus it cannot be recommended.

There are no general guidelines for the usage of evening primrose oil because it is a nutritional supplement. Adults are generally regarded safe with a daily dose of 500 milligrams, while many can tolerate up to 1,300 mg without experiencing any negative effects. Evening primrose should not be given to children without first consulting a healthcare expert due to a lack of research. Evening primrose oil is commonly sold in gel cap form and can be found in many health foods stores or pharmacies. Primrose oil in a bottle is also available, although it is more difficult to dose precisely.

Figure 25: Oenothera Biennis is the scientific name for the evening primrose plant. They are members of the Onagraceae family. Other names for it include German rampion, hogweed, fever-plant, and Kind's cure-all. The plant is edible and has been used for a variety of health advantages over time. Salads are garnished with the delicious blooms, while the immature seedpods are cooked. The leaves are then boiled. The plant is mostly found in North America's central and eastern regions.

Source: https://www.lybrate.com/topic/evening-primrose-oil-benefits-and-side-effects

F

Ficus Septica

Ficus septica is a tree or a shrub with pendulous branches that can reach a height of 3 to 25 meters. The bole can have a diameter of up to 30cm. The tree is gathered in the wild for therapeutic purposes in the area. When regenerating woods, it can be utilized as a pioneer plant. In the wild, it can be found in a variety of soil types. Fig trees have an unusual method of fertilization, with each species dependent on a single, extremely specialized wasp species that is completely reliant on that fig species to reproduce. Male, long-styled female, and short-styled female flowers, often known as gall flowers, are produced by the trees. All three types of flowers are included within what we commonly refer to as the fruit structure.

Certain types of herpes and wounds caused by poisonous fish are treated with latex. Purging is made from the latex of the leaves and fruits. The leaves are used to alleviate headaches as a sudorific. Previously, the dried leaves were used as an opium replacement or combined with it. External application of the leaves is reported to be anti-rheumatic. Colds, coughs, fevers, and fungal and bacterial infections are all treated with them. To treat dysentery or diarrhea, root scrapings or leaves have been combined with water and drunk (Salim et al., 2019). The roots are antiemetic, antidote, anti-asthmatic, and diuretic. To cure urinary tract infections, the crushed root is combined with coconut water and consumed daily. The roots are applied to boils as a poultice. The fruits are diarrheic. This species' stem bark was effective against tumor cell cultures. Tylophorine and tylocrebrine are the primary alkaloids. In comparison to the aqueous free extract, the ethanol-free extract has a lower impact. The LD50 of the fruit extract *Ficus septica* is greater than the maximum dose and is rather safe when given orally to rats, but it is not safe when used chronically, according to acute toxicity research.

Fennel

Also called *Foeniculum piperitum, Foeniculum Vulgare, Foeniculum Vulgare* Fruit, Garden Fennel, *Anethum Foeniculum, Anethum piperitum*, Bari-Sanuf, Bitter Fennel, Carosella and Fennel seed, Fennel is a fragrant, perennial herb with yellow blooms. It's originally from the Mediterranean, but it's now prevalent all around the world. As an anise-flavored spice, dried fennel seeds are commonly used in cooking. But do not mistake fennel with anise; they're not the same thing, even if they look and taste alike. The dried mature seeds and oil of the fennel plant are used to create medication.

Fennel is used to treating a variety of digestive issues, including indigestion, gas, bloating, loss of appetite, and newborn colic. It's also used to treat a cough with phlegm or pus, coughs, bronchitis, cholera, backaches, bedwetting, and vision issues. Fennel is used by some women to increase the flow of breast milk, promote menstruation, make the birthing process easier, and increase sex drive. Snakebites are treated using fennel powder as a poultice. Fennel oil is often utilized as a flavoring component in foods and beverages. Fennel oil can be used as a flavoring factor in some laxatives and as a fragrant ingredient in soap and cosmetics in other production processes.

When consumed by mouth in the proportions frequently seen in food, fennel is LIKELY SAFE. For the most part, there is insufficient evidence to tell whether it is safe for adults when used in therapeutic doses. Fennel can cause allergic skin responses in certain persons. People who are allergic to mugwort, celery, or carrots, are more likely to be allergic to fennel as well. Fennel can also make skin more sensitive to sunlight, making sunburn more likely. If you have fair skin, use sunscreen. Estrogen is present in several birth control medications. Some of the effects of fennel may be like those of estrogen. Fennel, on the other hand, is not as potent as the estrogen included in birth control tablets. When taken with birth control pills, fennel may reduce the effectiveness of the contraceptives. Use a supplementary type of birth control, such as a condom, if you are taking birth control tablets and fennel at the same time. In scientific studies, the relevant doses have been investigated: Note: A special multi-ingredient medication containing 164 mg of fennel, 97 mg of lemon balm, and 178 mg of German chamomile (ColiMil) twice daily for a week is recommended for colic in breast-fed newborns.

Ferula Assa-foetida

The dried fluid collected from the roots of Ferula plants is known as asafoetida (*Ferula asafoetida*). Although it is native to Afghanistan and Iraq, asafoetida is widely used in Indian cuisine, where it is dried and powdered into a spice known as hing. Aside from flavoring food, asafoetida has been used for millennia for its purported health benefits all throughout the world (Schultz et al., 2020). Asafoetida, which is technically a gum-resin, is a hard material taken from the Ferula plants' enormous, carrot-shaped roots. It's usually dried, pounded into a coarse, yellow powder, and utilized for culinary or medicinal reasons once it's been removed. Because of its high concentration of sulfur compounds, asafoetida has a strong, pungent odor as a spice. This seasoning is commonly referred to as stinking gum because of its foul odor. When cooked, however, its flavor and scent become considerably more appetizing, and it is frequently compared to leeks, garlic, and even meat.

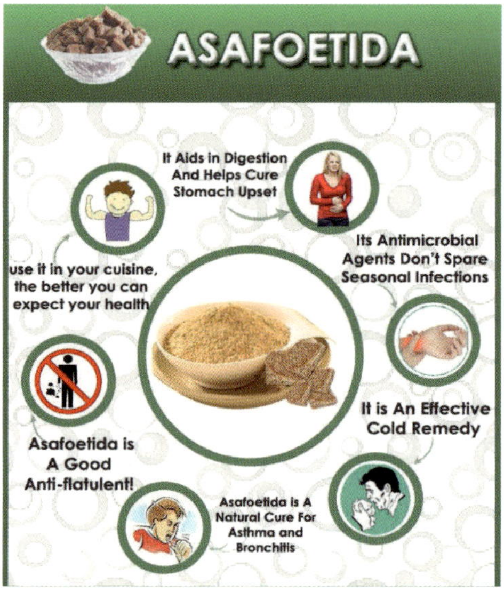

Figure 26: Weak digestion, antispasmodic, intestinal parasites, bronchitis, flatulence, stomachache, epilepsy, ulcer, asthma, whooping cough, and influenza

are among the disorders for which asafoetida is traditionally used. Asafoetida is a helpful treatment for a variety of stomach ailments.

Source: https://www.pinterest.com/pin/249105423118567990/

Asafoetida has been employed in traditional medicine for generations, in addition to imparting a particular flavor to dishes. Hing, for instance, is used in Ayurvedic medicine to help with digestion and gas, as well as to manage bronchitis and kidney stones. The dried gum was wrapped around the neck to try to fend off infections and sickness during the Middle Ages. Despite being utilized for countless generations, modern science has yet to confirm many of the traditional applications of asafoetida.

While there is little study on the safety of asafoetida in humans, the levels used in cooking are regarded to be safe for most people. The participants in one human trial showed that taking 250 mg twice a day for 30 days was well tolerated. Large doses of asafoetida, on the other hand, have been linked to tongue swelling, gas, diarrhea, nervousness, and headaches in animals. Furthermore, a mouse study indicated that doses larger than 455 mg per pound (1,000 mg per kg) of body weight may be hazardous (Saslis-Lagoudakis et al., 2011).

Asafoetida is also not suggested for pregnant or breastfeeding women, or young children, due to a lack of studies. People on blood pressure meds or blood thinning therapies should avoid asafoetida supplements since it can reduce blood pressure or thin the blood. Asafoetida is commonly combined with wheat or rice flour when used as a spice. As a result, items containing asafoetida (or hing) may not be gluten-free. When dining out at a restaurant that employs hing powder in its dishes, this can be a major worry.

Fumaria Officinalis

Also called Fumiterry, Fumus, Hedge Fumitory, Herba Fumariae, Beggary, Earth Smoke, Fleur de Terre, Fumaria, Fumaria officinalis, Herbe à la Veuve, Vapor, and Wax Dolls, Fumitory is a short shrub with gray pointy leaves that have the wispy appearance of smoke from a distance. It was given the term "earth smoke" as a result of this. Medicine is made from the portions of the plant that grow above ground.

Fumitory is used to alleviate gastrointestinal spasms and inflammatory bowel disease (IBD), as well as to stimulate the flow of bile, a fluid produced by the liver that is essential for fat digestion. Skin diseases, eye irritation (conjunctivitis), heart difficulties, water retention, and constipation are all treated with fumitory. Fumitory contains a chemical that may help to relieve bile duct or intestinal spasms.

When taken by mouth in the appropriate dosages for a short length of time, fumitory is POSSIBLY SAFE for most adults. However, consuming significant amounts of fumitory by mouth could be dangerous. Trembling, convulsions, and death are all possible side effects. Use only fumitory eye drop preparations that have been commercially sterilized. If the drops contain germs, you can acquire an eye infection. In breastfeeding and pregnancy, there is not enough trustworthy information on the safety of fumitory use when pregnant or nursing. To be on the safe side, avoid using it.

The proper dose of fumitory is determined by several criteria, including the user's age, health, and other circumstances. There is currently insufficient scientific data to identify an appropriate dose range for fumitory. Bear in mind that organic ingredients are sometimes not safe, and that dosages are crucial. Before using, be sure to read the label on the bottle and consult your pharmacy, physician, or other healthcare expert (Reinaldo et al., 2020).

Forsythia Suspensa

Also called Mimosa de Paris, Rengyo, Syringa suspensa, Weeping Golden Bell, Forsitia, Forsythia de Paris, *Forsythia Fructus, Forsythia koreana*, and *Forsythia, Forsythia suspensea* is a flowering plant. The fruit has medicinal properties. Bronchitis, fever, sore throat, heart illness, vomiting, tonsillitis, HIV/AIDS, gonorrhea, discomfort and swelling (inflammation), and a severe skin rash with vomiting and fever caused by bacteria are all treated with forsythia (erysipelas).

For the treatment of bronchiolitis, forsythia is sometimes given intravenously (IV) in combination with other herbs. Forsythia has been shown to reduce inflammation. More information is required, however, to determine how forsythia works. It is unknown whether forsythia is safe to consume. There is some evidence that an injectable version can be used safely in children.

Figure 27: According to studies, fumitory can aid with gallbladder disease, liver inflammation, and bile flow from the gallbladder. It has been demonstrated to increase liver function by up to 80%, as well as detox and heal the liver in general... as well as boosting liver activity. Fumitory also works as an antispasmodic for the gallbladder and reduces the production of gallbladder stones.

Source: http://shaafiherbalcare.com/blog/fumaria-officinalis/

The right dose of forsythia is determined by a number of factors, including the user's age, health, and other circumstances. There is currently insufficient scientific evidence to define an optimal dosing range for forsythia. Bare in mind that organic ingredients are sometimes not safe, and that dosages are crucial. Before using, be sure to read the label on the bottle and contact your pharmacist, physician, or other healthcare expert.

Fucus Vesiculosis

Also called Sea Oak, Seawrack, Varech, Varech Vésiculeux, Black Tang, Bladder Fucus, Bladder Wrack, and Kelp, *Fucus vesiculosis* is a brown seaweed kind. The entire plant is used to manufacture medication. Thyroid diseases, iodine deficiency, obesity, and a variety of other ailments are all treated with Fucus vesiculosus, however, there is no clear scientific evidence to back up these claims. It is also possible that using Fucus vesiculosus is dangerous. Iodine is present in different levels in Fucus vesiculosus. Iodine may aid in the prevention or treatment of thyroid diseases. Fucus vesiculosus may also have anti-diabetic properties and influence hormone levels. However, further data is required. When administered to the skin, Fucus vesiculosus is possibly safe. When eaten by mouth, it is possibly unsafe (Rawat et al., 2020). It could contain excessive levels of iodine. Iodine in excess can cause or exacerbate thyroid disorders. Heavy metals, which can induce heavy metal toxicity, may be present. Thyroid disorders should not be treated without first consulting a doctor.

The right dose of Fucus vesiculosus is determined by a number of factors, including the user's age, health, and other circumstances. There is currently insufficient scientific evidence to define an acceptable dosing range for Fucus vesiculosus.

Foxglove Digitalis

Also called Purple Foxglove, Scotch Mercury, Throatwort, Witch's Bells, Woolly Foxglove, Digitale, Digitale Laineuse, Digitale Pourpre, Fairy Finger, and Foxglove, Foxglove digitalis is a type of plant. Although the sections of the plant that sprout above soil can be used for medication, self-medication with foxglove is not recommended. The plant is harmful in all sections. A prescription medicine called digoxin is made from chemicals extracted from foxglove. In the United States, *Digitalis lanata* is the most common source of digoxin. Foxglove is often used to treat heart failure, liquid build-up in the body (congestive heart failure or CHF), and irregular heartbeat (atrial fibrillation). However, it is not suitable for any usage.

Foxglove includes compounds that are used to make the pharmaceutical drug digoxin (Lanoxin). These substances can make cardiac muscle contractions stronger, modify heart rate, and enhance heart blood flow. By mouth, foxglove can help with abnormal heart rhythms including atrial fibrillation and flutter. However, using foxglove for this ailment without the guidance and supervision of a healthcare practitioner is risky. By oral, foxglove may help with CHF and CHF-related edema. However, using foxglove for this ailment without the guidance and supervision of a healthcare practitioner is risky.

Without the advice and supervision of a healthcare professional, taking foxglove by mouth is extremely dangerous. Some persons are particularly vulnerable to the harmful effects of foxglove and should avoid it at all costs. Foxglove can lead to cardiac problems and perhaps death. Stomach upset, vomiting, nausea, strong sluggish pulse, blurred vision, tiny pupils, dizziness, excessive urine, exhaustion, tremors and muscle weakness, stupor, convulsions, disorientation, irregular heartbeats, and death are all symptoms of foxglove poisoning. Long-term foxglove use can cause toxicity symptoms such as visual halos, yellow-green eyesight, and stomach trouble.

The right dose of foxglove is determined by a number of factors, including the user's age, health, and other circumstances. There is currently insufficient scientific data to define an appropriate dosing range for foxglove.

Forskohlii, Coleus

Also called Colforsine, Forskohlii, Forskolin, Forskolina, Forskoline, Coleus, Coleus Barbatus, Coleus Forskolii, Coleus Forskohlii, and Coleus Penzigii, Coleus is a plant that has been utilized to treat cardiac problems including high blood pressure and chest pain (angina), as well as respiratory problems like asthma, since ancient times. The chemical forskolin is found in the roots of the coleus plant. Coleus is used to treat dry eyes, allergies, skin conditions like psoriasis and eczema, obesity, painful menstrual periods, urinary tract infections (UTI), irritable bowel syndrome (IBS), bladder infections, blood clots, advanced cancer, insomnia, sexual problems in men, and convulsions when taken by mouth (Peng et al., 2017).

Coleus is sometimes given intravenously (IV) to patients with heart failure. Coleus powder is inhaled by some people who suffer from asthma. Coleus drops are used to treat glaucoma in the eyes. Coleus extracts with significant quantities of forskolin are now being produced by herbal product makers. These supplements are being advertised for the same ailments that forskolin has been utilized for in the past. However, there is currently no clear scientific evidence that Coleus extracts taken orally are effective.

When taken orally, intravenously (IV), inhaled (breathed in), or administered as ocular drops, Coleus is potentially harmless for most individuals. Nevertheless, there may be some negative consequences. Coleus can produce flushing and low blood pressure when administered by IV. Coleus can irritate the throat, produce coughing, tremors, and restlessness when inhaled. Coleus eye drops can irritate the eyes and cause stinging.

G

Garlic

Figure 28: Garlic can be grown in both large scale and small scale farms.

Source: https://savvygardening.com/how-to-grow-garlic-in-pots/

Garlic is commonly referred to as Allium sativum. It belongs to the genus Allium. It is a perennial bulb with a flowering system that can grow up to 1m. A garlic plant is known to have a flat leaf blade with an acute apex. Its leaf blade is also linear and can be 1.3 to 2.5 cm wide. When flowering, garlic plants produce flowers purple or pink in color. Garlic bulbs usually have outer layers composed of thin sheathing leaves. The outer layers surround an inner sheath which surrounds the clove. The cloves are usually symmetrical in shape in exception of those that are close to the center. Usually, a garlic bulb contain up to 20 cloves. Though it is mostly used in favoring foods, it is also an important medicinal plant. Garlic was heavily used in traditional medicine (Ningthoujamet et al., 2012). As its use dates to the early century, it is grown and used extensively globally. Extensive usage of garlic is due to the remarkable medicinal usage. Garlic is known to have preventive characteristics. Fresh garlic is a good source of minerals, vitamins and trace elements though they are found in small amounts. Among the trace elements in garlic include selenium and germanium. The two elements contribute to garlic's antitumor effect. Garlic's chemical composition also contributes to its medicinal value. Garlic is composed of a volatile oil having sulfur-containing compounds such as methylallytrisulphide, diallyltrisulfide and diallyldisulfide. The volatile oil also contains ajoenes, vinyldithiins and alliin which is a sulfur-containing amino acid. Garlic can be consumed in its natural form. However in some cases they are consumed as health supplements. Health supplements usually depending on the amount of allicin present.

Synonyms

There are a variety of words that could be used to mean garlic. They include ail, seasoning, onion, favorer, Allium sativum, flavoring, leek, alliaceous plant, seasoner and flavourer. The most commonly used synonym is Allium sativum.

Usage

Garlic can be used to treat a variety of diseases more so those related to the heart and the blood system. Among the common uses of garlic is preventing occurrence of common cold adults. Continuous consumption of garlic is said to reduce the frequency of the occurrence of common cold. Cooked and raw garlic are beneficial. However, better results are achieved when garlic is eaten raw.

Figure 29: There are a variety of garlic oriented home remedies used in treating a number of diseases.

Source: https://cdn-prod.medicalnewstoday.com/content/images/articles/265/265853/bulbs-and-bowl-of-garlic.jpg

Over the years garlic has been used by women when dealing with issues of vulvovaginal candidiasis. The treatment lasts for 14 days and involves the consumption of garlic powder. It is useful in dealing with other menstrual disorders. Garlic has also been used in dealing bacterial infection as it is a good antimicrobial. It has also been used in treating a number of cancer cases. Garlic is said to lower the risk of lung cancer. Studies show that raw garlic can be used as a chemo-preventive agent for lung cancer. Garlic has also been used in dealing with brain cancer (Nadda et al., 2020). The organo-sulfur compounds in garlic can destroy cells in glioblastomas. The sulfur compounds are very effective in eradicating brain cancer cells. Garlic has also been used in preventing the risk of prostate cancer, stomach cancer, colon cancer, breast cancer and rectal cancer.

Figure 30: Garlic contains heart-protecting elements.

Source: https://cdn-prod.medicalnewstoday.com/content/images/articles/265/265853/garlic-in-heart-shaped-bowl.jpg

Garlic has greatly been used in heart protection. Garlic contains a component known as Diallyl trisulphide which is said to protect the heart after a heart attack and during cardiac surgery. There is a possibility that it could be used in treating heart failure and protecting the heart from diabetes-induced cardiomyopathy. Patients suffering from ischemic heart disease also use garlic in treating the disease. Garlic has proven to be useful in improving endothelial function and vascular elasticity. Garlic is also used in dealing with issues of high cholesterol as well as high blood pressure. Patients suffering from high blood pressure and high cholesterol are usually given garlic extract supplements which help improve their blood lipid profile, reduce diastolic and systolic blood pressure and improve blood antioxidant ability. Garlic extract supplements also help lower the level of oxidation products in the blood.

Garlic is also useful in reducing alcohol-induced liver injury. This kind of injury is caused by the long-term consumption of alcohol. Diallyl disulfide in garlic helps minimize the effects of ethanol on the liver preventing oxidative stress. Women use garlic to reduce the risks of hip osteoarthritis. Garlic is rich in allium which decreases levels of osteoarthritis. Other compounds found in garlic have been said to help treat the condition. Garlic is also used in treating peripheral vascular disease. Peripheral vascular disease is a circulatory condition where blood supply is limited to the limbs as a result of narrowed blood vessels. It is an indication of fatty deposits on the arteries. Garlic is used in improving blood circulation allowing pain-free walks (Miara et al., 2018).

Garlic is used among patients who suffer from tinnitus. Tinnitus is a disease characterized by ringing in both ears and is commonly linked to hearing loss. For effective treatment, garlic is used alongside other supplements. However, garlic is not recommended for patients who suffer from chronic cases of tinnitus. Garlic is also used in treating ear infections among other infections. Garlic is considered a powerful antibiotic. It is more powerful than other antibiotics. As an antibiotic, garlic is mostly used in treating intestinal infections.

Pregnant women consume large amounts of garlic. This is done to avoid pre-term delivery. During pregnancy, pregnant women are very vulnerable in that any microbial infection could cause premature delivery. Garlic contains antimicrobials and is useful in lowering the risk of microbial infection, therefore, reducing the risk of premature pregnancy. Other than chronic illnesses, garlic is used in dealing sinus congestion, fever, hemorrhoids, coughs, asthma, headache, diarrhea, stomach ache, bloody urine, low blood sugar, bronchitis, scalp ringworm, tooth sensitivity, stomach inflammation, tuberculosis, whooping cough and shortness of breath. Others use garlic to relieve fatigue and stress

Dosage

Garlic dosage varies according to the components needed and the availability of numerous products. Dosage also varies depending on the administration. There are two main methods of dosing: by mouth and skin application.

Oral dosage (by mouth)

To harden the arteries: A garlic powder tablet is administered. The tablet is taken three times in a day or as a single dose. The dosage is taken for 4 years. Also, 150 mg of certain garlic supplements are taken by patients. The supplements are taken twice daily for two years. In other cases, the patient is given a combination of products having certain amounts of garlic. For others, a garlic extract supplement has 250 mg of aged garlic extract is administered daily for a year. For other patients, their treatment involves the use of a combination of products having about 300mg of aged garlic extract. In this case, the dosage is four tablets taken daily for a year.

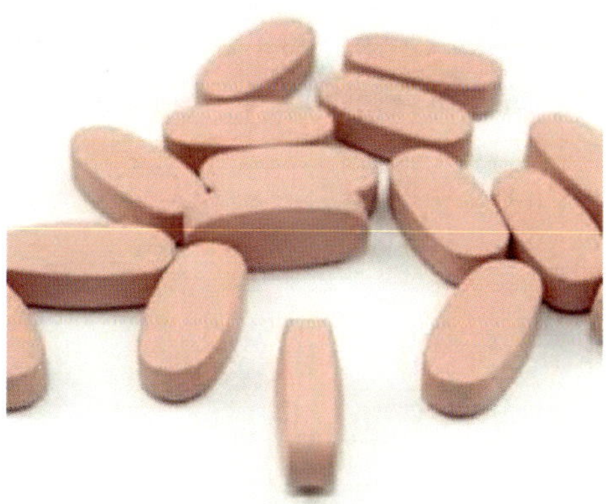

Figure 31: Garlic is processed to produce tablets and capsules.

Source: https://m.made-in-china.com/product/GMP-Certificated-Garlic-Powder-Tablet-Garlic-Softgel-784861192.html

For rectal and colon cancer: Capsules are having about 2.4 ml of aged garlic extract are administered. The capsules are taken daily for two years.

To treat high cholesterol: A dosage of about 1000 mg to 7200 mg of a given garlic extract is administered to patients. The doses are divided and taken daily for 4-6 months. A garlic powder tablet of 600 mg to 900 mg is administered daily. The dosage could be divided into two or more doses. There is another garlic powder tablet of 300 mg is also administered. The tablet is taken twice in a day for 12 weeks. In other cases treatment involves patients taking 3 grams of fish oil in addition to 1200 mg of garlic powder. The treatment lasted for 4 weeks. In other cases, 600 mg of fish oil and 500 mg of garlic oil is administered daily for 60 days (Menendez-Baceta et al., 2015).

To treat high blood pressure: Garlic tablets of 300 mg-1500 mg are administered in divided doses daily. This treatment lasts for 24 weeks. Single doses of a certain powder tablet of 2400 mg. In other cases, it is taken for 12 weeks when 600 mg of the dosage is taken. For other patients capsules have 960 mg-7.2 grams of an aged garlic extract. The dosage is divided into three and taken daily for half a year. Treatment may involve the administration of certain products having given amounts of aged garlic extract. Some of these products include Kyolic. 600 mg of fish oil and 500 mg of garlic oil can also be administered for 60 days.

For tick bites: Medication involves the administration of capsules having 1200 mg of garlic. The dose is taken daily for 8 weeks.

For prostate cancer: Dosage is the administration of a water-soluble garlic extract. It is taken daily for one month.

Skin application

For skin application, the dosage is mostly used to treat fungal infections. The dosage involves garlic ingredient known as ajoene is applied twice a day for a week.

Possible side effects

Garlic can safely be consumed by a large number of the world's population. However, it has some side effects such as heartburn, bad breath, diarrhea and gas when consumed by mouth. The effects are worse when raw garlic was consumed. In some cases, garlic is said to cause allergic reactions while in others it increases bleeding risks and clotting time. When applied to the skin, there are possibilities of garlic causing skin damage bearing similarities to a burn. This is the case is raw garlic was used. In severe cases, raw garlic causes severe skin irritation.

Extracts

Gardenia Gummifera

Figure 32: *Gardenia gummifera* plants produce resins used in treating different diseases

Source: https://encrypted-tbn0.gstatic.com/images?q=tbn:ANd9GcQhqOwqos xB7WshkqmTNO6vi4agE3zU_xmSCg&usqp=CAU

Gardenia gummifera is a tree commonly grown in India. It is of the family Rubiaceae which is the coffee family. Other names for the plant are Genipa arborea Baill and Genipa gummifera Baill. The tree is relatively small and can grow to a height of 15 m tall. They are evergreen and have leaves in the opposite. In other cases, the leaves are in whorls of 3-4. The leaves have a glossy with a leathery texture and are also dark green in color. Their leaf blade can either be obovate-oblong, obovate or ovate (Muchugi et al., 2006). The base of the leaf blade is usually blunt. In other cases, it is heart-shaped or round. The tips of the leaves are pointed or blunt. When flowering, the tree produces flowers white or pale yellow. Flowers have about 5 to 12 petals. The tree produces oblong berries which are green when the plant is young. The fruit also produces a lot of seeds.

The tree is known for its herbal properties. It is used for its medicinal value. Gardenia gummifera is used in making a paste that is applied over areas with pain or wounds. Once the paste is applied, the area is tied with a clean bandage to minimize complaints. Gardenia gummifera produces resins that are kept over the gum or tooth affected by inflammation or dental carries. Doses having 200 mg of resin is dissolved in buttermilk or water and administered to patients suffering from appetite loss or indigestion. Resin from Gardenia gummifera is dries and dissolves in water. The mixture is then administered to patients suffering from piles, intestinal worms, constipation and distention of abdomen. A dose having 300-500 mg of Gardenia gummifera is used in treating fever brought about by indigestion. The plant is also used in treating coughs and dyspnea. This form of treatment involves resin being dissolved in warm water and the mixture is administered to patients. 500 mg of the plant is used in treating skin diseases.

Even though the plant is useful in treating some diseases. Necessary precautions are to be taken to avoid adverse reactions. One of the precautions is that resins of Gardenia gummifera should not be consumed on an empty stomach. On an empty stomach, the resin is said to cause nausea and hyperacidity.

Gossypium sp.

Figure 33: Properties of *Gossypium sp.* contribute to its medicinal value.

Source: https://en.wikipedia.org/wiki/Gossypium

Gossypium sp is commonly known as cotton. From this plant cotton is harvested. It is a flowering plant of the mallow family. It is commonly grown in tropical and subtropical areas. There are several words that can be used in the place of *Gossypium sp*. They include Sturtia R.Br., Notoxylinon Lewton, Ingenhouzia DC, Thuberia A.Gray and Selera Ulbr. The plant is usually composed of ribbed branchlets that are winged or angled. They are also gland-dotted. Their leaf lamina takes an orbicular outlook. Their leaves are lobbed in 3s or 5s. The leaf apex is acuminate. The leaves usually have 3-5 nerves and the petiole can measure 8 cm. The plant produces pale-yellow flowers having dark centers. Flowers can measure a diameter of 5 cm (Maruca et al., 2019).

Gossypium sp was used in traditional medicine. Some of the remedies are used today. Early research shows that *Gossypium sp* was used in the treatment of malaria. The treatment process involved the preparation of medication composed of Royal Poinciana, *Gossypium sp*, angular winter cherry and Barbados nuts. The medication is administered orally for 7 days. It was said to eradicate the malaria parasites among patients suffering from malaria. Other than treating malaria, *Gossypium sp* is used in treating diarrhea, fever, menstrual disorders, nausea, headache, male birth control, menopausal symptoms, abnormal bleeding, ovarian cancer and inducing labor. The advantage of using *Gossypium sp* as medicine is that it is safe for consumption. However, some side effects have been noted among men who use it as birth control. *Gossypium* can cause irreversible sterility. It is also unsafe for pregnant women as it could cause miscarriage. It is also not safe for people suffering from kidney and reproductive problems. Dosing of *Gossypium sp* varies depending on several factors such as underlying health conditions and age. The method of consumption of the dosage. It can be ingested as supplements or a home remedy.

Gmelina Arborea

Figure 34: *Gmelina arborea* is native to the Indian sub-continent.

Source: https://keyserver.lucidcentral.org/weeds/data/media/Html/gmelina_arborea.htm

Gmelina arborea commonly known as Gamhar or Gambhari is a tree of the lamiaceae. It is known to be a bitten tonic. Traditionally, the leaves, roots and fruits of the plant were used in treating many diseases. It was used as a diuretic, anthelmintic, anti-aging, hepatoprotective, analgesic and anti-diabetic agent. The tree is said to take a short time to grow. It thrives in moist fertile soils and cannot thrive in dry areas. A *Gmelina arborea* tree can grow to a height of 30m. When fresh, its wood is pale-yellow to a pinkish-buff and the tree turns yellowish-brown when exposed to sunlight (Martínez & Luján, 2011). When flowering, the plant loses a significant number of leaves. The tree produces fruits. The fruits of *Gmelina arborea* are edible. The fruit is usually smooth and dark green in color. It can reach up to a length of 2.5cm and is yellow when ripe. It also has a fruity smell and a bitter-sweet taste. The entire plant *Gmelina arborea* plant is used in medicine. The chemical composition of the different parts of the plant allow it be used to treat various diseases.

Some of the uses of *Gmelina arborea* include it being used to reduce inflammation and reduce pain. It is also used to improve indigestion as well as absorption of food in the body. *Gmelina arborea* is said to help ensure all activities in the heart run smoothly. It is also useful in increasing brain functions and to some extent, it can improve intelligence. It is used among both men and women as a natural anti-aging agent and an aphrodisiac agent. As an herb, Gmelina arborea helps prevent hemorrhages. Lactating mothers use *Gmelina arborea* to improve milk production. It is also used by men to improve the quality and quantity of semen. It is also useful in promoting hair growth and nourishing hair. It is an essential natural agent useful in dealing with bleeding disorders, burning sensations and stabilizing abnormally raised temperatures.

Gmelina arborea fruits are used in treating tuberculosis as it helps fasten cavitation in the lung. The plant is also used to make a paste which is applied on the forehead to lower fever and deal with headaches. Its roots are used in treating ascites and when decocted it is used in healing abdominal tumors. The leaves of *Gmelina arborea* have been used in treating leprosy. Other than the root, the bark of *Gmelina arborea* is used as an agent to deal with fever, urinary infection, hallucination, abdominal pain, piles and tridoshas. Dosages are mostly administered by mouth. They could be in form of concoctions or fruit powder. The plant has also been used to make other medicinal products. *Gmelina arborea* is safe for consumption with minimum cases of side effects being recorded. In cases where side effects occur, they tend to be severe. Many cases of side effects are also recorded when the plant is used alongside other medication.

Glycyrrhetinic Acid

Glycyrrhetinic acid is obtained from Glycrrhizic acid. The Glycrrhizic acid is derived from a licorice plant. The roots are used in the making of this acid. The acid contains both biological and pharmacological activities that are useful in the medical field. Among the use of Glycyrrhetinic acid is that it is used in the treatment of eczema (Mati & de Boer, 2011). After application, the medicine is said to reduce swelling, redness and itching. It is also used in controlling canker sores inside the mouth by reducing the pain and fastening the healing process. It is also useful in among patients who have undergone breathing tube removal. It deals with the side effects of the process. The acid is used in making drugs that are administered to patients. It is safe for patients provides they follow the dosage as per the physician's prescription. There are some side effects noted when the drugs are consumed in large quantities. Some of the side effects include weakness and low potassium levels.

Gamma-amino-butyric Acid

Gamma-amino-butyric acid (GABA) is an acid produced in the body and heavily used by the brain. It is used as a neurotransmitter and plays the role of a chemical messenger. GABA can inhibit or block certain brain cells thereby lowering activity in the nervous system. In as much as GABA is heavily utilized by the body, few foods provide the body with this acid. In most cases, fermented foods are said to have high levels of GABA. One of the major ways GABA is supplied to the body is through supplements. Some of the uses of GABA include it being used in calming the brain. GABA tends to leave a calming effect on the brain. This has led to a significant number of people using GABA to reduce stress. GABA supplements are also used to help people with mood disorders such as depression, seizure disorders, attention deficit hyperactivity disorder, movement disorder, panic disorder and anxiety. GABA is also used among people suffering from high blood pressure and insomnia among other conditions.

Dosage is administered orally in the form of supplements. Some of the recorded side effects of GABA supplements include muscle weakness, upset stomach, sleepiness and headache.

Garcinia Cambogia

Garcinia cambogia is a plant native to India and Southeast Asia. The plant produces a fruit that contains a chemical used in making medicine. It is used among adults to help improve their athletic performance. Chemicals found in the fruit boost adults' ability to exercise. *Garcinia cambogia* helps reduce obesity as it initiates weight loss to a certain degree. It also treats joint pain, constipation, diarrhea and schistosomiasis among other conditions. Dosage is usually administered by mouth. The dosage used for different conditions is influenced by several factors such as the underlying conditions of the user, their health and age. As it is a natural product, dosage should be administered according to the prescription of the physician. Some of the possible side effects of Garcinia cambogia is live problems. Mild side effects include headache, nausea and discomfort in the digestive tract (Li et al., 2015).

Gastroesophageal Reflux Disease

Gastroesophageal reflux disease is also known as GERD is a disorder that affects the digestive system. The disease affects the rings of muscle found between the stomach and the esophagus. The affected region is at times referred to as the lower esophageal sphincter.

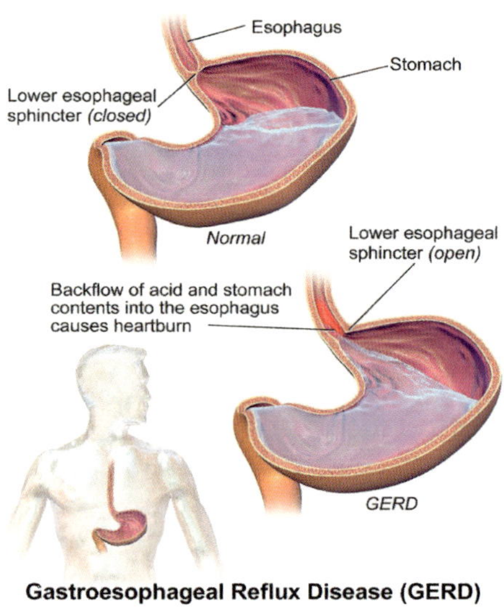

Figure 35: GERD is linked to a condition called hiatal hernia.

Source: https://en.wikipedia.org/wiki/Gastroesophageal_reflux_disease

The disease causes acid indigestion and heartburn. The condition involves food and other substances found in the stomach going back to the esophagus. GERD occurs when the lower esophageal sphincter is weak or when it relaxes in times that it shouldn't. In normal cases, the LES opens up to allow food to enter the stomach and closes to prevent stomach juices from going to the esophagus. Some of the risk factors of GERD include smoking, obesity, pregnancy, certain foods and drinks, gastroparesis and specific medications. Heartburn is said to be a common symptom of GERD. Other symptoms include vomiting, bad breath, worn out enamels and nausea. Treatment is mostly done to reduce reflux and protect the esophagus lining from damage. Treatment involves the administration of antacids, proton pimp inhibitors, prokinetics and H2 blockers. Patients may also be advised to make some changes to their lifestyle. In severe cases of GERD patients may be required to undergo surgery.

Green Farmacy Garden

The green pharmacy garden is also known as the healing garden. It was brought to existence by Dr. Duke. He was a botanist. The garden is said to contain 300 plants used in the treatment of different kinds of diseases and is used in herbal medicine. The establishment of the garden was done to create cue cards utilized to prompting conversations (López-Malo et al., 2020). Plants such as garlic were grown to help treat colds and lower cholesterol levels. Turmeric was grown for its antioxidant properties while the European stinging nestle was grown to lower the pain of osteoarthritis. Duke was able to develop several remedies that can be used for different diseases. From his discoveries, he has written several books among them "The Green Pharmacy". The general public is allowed to visit the garden and learn more about the herbs.

Gamma-linoleic Acid

Gamma-linoleic acid is found in several plant seeds and is used as medicine. It is used in dealing with skin problems such as eczema, systemic sclerosis and psoriasis. It is also used in the treatment of rheumatoid arthritis, diabetic nerve pain, high cholesterol, and metabolic syndrome, polyps in the mouth, depression and heart disease. It is used among breast cancer patients to improve response to treatment drugs and also prevent cancer. Gamma-linoleic acid is safe for consumption for taken properly. Some of the side effects include belching, soft stools, intestinal gas, and diarrhea. It lengthens the time taken for blood to clot making it a concern during surgery.

Genetically Modified Organism

Genetically modified organism has undergone genetic modification by the use of genetic engineering. The result of the process is an organism bearing certain genes. A genetically modified organism could be a plant, animal or fish. Genetic modification has become popular over the years because it can help improve food security globally. More advancements are being made on the kind of technology used during generic modification so that the results are as required. Animals that are genetically modified organisms are used in laboratories in testing drugs and the prevalence of diseases under certain conditions. They are used in studying how certain genes are linked to diseases. Other genetically modified organisms are used for human consumption. The advantages of genetically modified organisms used for human consumption is that they mature faster, have larger yields, reduced incidence to diseases, have good taste and longer shelf life.

Guinea Pig

Guinea pigs are animals in the rodent family kept as pets and used in scientific research. It has a robust body with short ears, short limbs and large eyes and head. Their feet have short sharp claws and hairless soles. They have three toes on the hind feet and four on the forefeet. There are different breeds of guinea pigs suited to survive the environmental conditions. Guinea pigs can be distinguished by their length and coat texture (Kujawska et al., 2017). Their coat color varies from black, white, brown, tan, reddish and cream. Guinea pig are herbivores and need a significant amount of water in their diet. Female guinea pigs can give birth to about 13 young ones. A significant number of guinea pigs are used in laboratories for research purposes. They are used as test subjects to check the effectiveness of medicine and the relation between certain diseases and specific factors.

Glucosyltransferase

Glucosyltransferase is an enzyme (cell-associated or extracellular) produced by streptococci. It is utilized in the biosynthesis of polysaccharides used outside cells. Glucosyltransferase can also be synthesized by S.sanguis. S.sanguis and S. mutans easily bind with one another and bacteria in the presence of Glucan. Dental plaques create a conducive environment for the formation of Glucan. Glucosyltransferase has been identified as a possible antigen that can be used in vaccines such as the human caries vaccine. For this reason, it is used for oral immunization.

H

Helianthus Annus

Figure 36: Helianthus annus is among the different types of helianthus.

Source: https://www.gardenia.net/plant/helianthus-annuus

Helianthus annus commonly known as the common sunflower is a plant known to produce both edible fruits and oil. In some countries, it is utilized as wild bird fruit. The plant was originally wild and was later on domesticated in Americas. The wild variety of *Helianthus annus* is known to have several flower heads. This is very different from the domesticated *Helianthus annus* which has one flower head. Its name sunflower was as a result of the appearance of the flower head which bears similarities to the sun. Americans introduced sunflower seeds to Europe. From there it was processed to produce sunflower oil which is a common cooking ingredient. The Helianthus annus has a straight hair stem which is rough. It can grow up to a height of 3 m. The tallest sunflower recorded a height of 9.17 m. The leaves of a *H. annus* are rough, broad, alternate and roughly toothed. In a *H. annus,* the part referred to as the flower is the flower head. The flower head is made up of florets (Kunwar et al., 2016). Florets are the individual petaled flowers within a flower. In the H. annus, the outer flower are referred to as ray flowers. Each of the petals in the flower has a ligule made up of fused petals. The petals can either be orange, yellow or red and are sexually sterile. Flowers found at the center of the flower head are called disk flowers. Flowers in these regions develop and become the seeds. These flowers have a spiral arrangement.

Synonyms

There are a variety of words that could be used in the place of *H. annus*. They include giant sunflower, tall sunflower, *H. angustifolius*, common sunflower, Girasol, mirasol, Jerusalem artichoke, *H. tuberosus*, *H. petiolaris*, and *H. laetiflorus*.

Usage

Helianthus annus has a broad area of usage. The most commonly known use of *Helianthus annus* is the production of sunflower oil to be used during cooking. Some of the medicinal uses of *H. annus* include its leaves being used to make tea which is considered an expectorant, astringent and diuretic. The tea is used in bringing down high fevers. To deal with cases of snake bites, sores, spider bites and swellings, the leaves of *H. annus* are crushed and used as a poultice. To allow long-term use of the leaves, they were dried and stored. The flowers of the plants also have medicinal value. They are used in making tea which when consumed is said to treat lung ailments and malaria (Kovach et al., 2009). Other than the leaves, the seeds of *H.*

annus are also expectorant and diuretic. They have successfully been used in the treatment of pulmonary issues. Both the seeds and the fruit head of the *H. annus* are stomachic, febrifuge and nutritive. The roots of the plant are used in making a decoction utilized as a warm wash on rheumatic pains and aches. It is also used as an alternative to other oils as it lowers cholesterol. Certain brands of sunflower oil is used in the treatment of athlete's foot. In other cases, *H. annus* is used in the hardening of arteries, tackling skin conditions and constipation.

Dosage

Several factors affect the type of dose given to an individual. They include underlying conditions, age and health. An appropriate range of dose for *H. annus* has not yet been established. For this reason patients are advised to use *H. annus* products as per the physician's prescription.

Possible side effects

There are no recorded side effects of *H. annus* when consumed orally or when applied to the skin in recommended amounts. For individuals who are allergic to plants from the Compositae family, there are likely to experience some allergic reactions.

Hibiscus Rosa-sinesis

Hibiscus rosa-sinesis are plants known to bloom continuously. They are also known as the Chinese Hibiscus. The blooming process results in the production of flowers of different shapes. Other common names for the *Hibiscus rosa-sinesis* are China rose, Chinese Hibiscus, Hawaiian Hibiscus and tropical hibiscus. The plant is evergreen and takes the form of a vase-shaped shrub. It is of medium texture and has a fast growth rate. It flowers and produces fruit throughout the year. There are different varieties of the *Hibiscus rosa-sinesis* and they can all be categorized according to the color of the flowers.

Figure 37: Hibiscus rosa-sinesis is native to China.

Source: http://www.plantsoftheworldonline.org/taxon/urn:lsid:ipni.org:names:560756-1

Medicinal uses of the *Hibiscus rosa-sinesis* include treatment of menstrual abnormalities, stomach aches, spasms, gonorrhea and bronchitis. It is also used in the pharmaceutical and therapeutic industry due to its composition. The plant is considered an antioxidant, anticonvulsant, aphrodisiac, anti-inflammatory, anxiolytic and antipyretic hence it is used in treating a variety of diseases.

Hordeum Sativum

Figure 38: Seeds of Hordeum sativum.

Source: http://theworldwidevegetables.weebly.com/hordeum-vulgare---barley.htmlhttp://theworldwidevegetables.weebly.com/hordeum-vulgare---barley.html

Hordeum sativum is commonly known as barley. It is an annual crop that can grow up to a height of 1m. There are different species of *Hordeum sativum* and have both female and male organs on the same plant. *Hordeum sativum* is wind-pollinated. Though the plant can survive in various soil types, it does preferably well in well-drained soil. The plant does not thrive in the shade. It does well in moist soils. The plant is relatively strong as it can withstand strong winds. Other names for *Hordeum sativum* are *Hordeum durum, Erumentum hordeum, Secale orientale* and *Hordeum nigrum*. The plant is grown for its edible seeds used as a salt, coffee, and sweetener and for making drinks. It also has medicinal value. Hordeum sativum is said to be an expectorant, stomachic, abortifacient, diuretic and febrifuge. Diuretic properties are attributed to the shoot while the seed sprouts are stomachic, expectorant, lenitive and galactofuge. In some instances, the roots are abortifacient (Karmakar et al., 2010). For this reason, the seed sprouts and shoots are used in treating dyspepsia, breast distension, infantile lacto-dyspepsia and regurgitation of milk. However, the treatment cannot be administered to lactating mothers as it could reduce the flow of milk.

Seeds of *Hordeum sativum* are stomachic, emollient, febrifuge, nutritive and digestive. It is consumed as highly nutritious food. In other cases, the germinated seed gets infused with water and the solution is mostly used among babies. The solution is said to help minimize excessive lactation. To help in the treatment of wounds and burns, *Hordeum sativum* is used in making a poultice. During its use in traditional medicine, *Hordeum sativum* was used for its antitumor activity. When developing a treatment for hepatitis, it was discovered that the germinating seed of *Hordeum sativum* creates a hypoglycaemic effect which is then followed by a hyperglycaemic activity. More research is underway to prove that *Hordeum sativum* can be used to handle cases of diabetes. *Hordeum sativum* is said to help reduce blood cholesterol levels. It is used in the prevention of bowel cancer and treatment of diarrhea and bronchitis. It is also used in weight loss.

Hyoscyamus Niger

Hyoscyamus niger is a plant known to be poisonous when consumed in large amounts. It is also called stinking nightshade, henbane or black henbane. *Hyoscyamus niger* is said to originate from Eurasia and belongs to the Salicaceae family. The plant is mostly used in the medical field more so in pharmaceuticals. Although the plant originates from Europe, it is rarely found in northern Europe. In eastern and central Europe has made the cultivation of Hyoscyamus niger for medicinal use is legal. The popular use of *Hyoscyamus niger* has led to the plant being listed as endangered species. Traditionally, Hyoscyamus niger was used in treating ailing bones, stomach pain, rheumatism, cough, toothache, nervous disease and asthma. In some communities, it was considered as a narcotic, analgesic or sedative. For people who have travel-sick problems, adhesive bandages fitted with henbane extract is put behind the ears. This helped reduce the discomfort. Both the roots and leaves of the Hyoscyamus niger is used as medicine.

Hyoscyamus niger was prepared by chopping the leaves and drying them. Once dried they would be used as medication, in making beer and tea, as incense or in the seasoning wine. Henbane oil is derived by boiling *Hyoscyamus niger* in oil. Incense blends are made using Hyoscyamus seeds. High levels of accuracy should be maintained as extremely high amounts of the drug could be toxic to the body. The plant also has therapeutic uses. A lethal dosage has not yet been established. Other than human beings, the plant is considered toxic to animals. This is in the exception of pigs as their bodies are immune to the herbal effects (Kayani et al., 2015). When consumed by human beings the plant causes peripheral inhibition followed by central stimulation. When consumed by human beings, some of the possible side effects include flushed skin, hallucinations, restlessness and dilated pupils. In certain cases, some of the effects include ataxia, vomiting, tachycardia, hyperpyrexia, convulsions and hypertension. In some cases, the side effects last for about four hours and in other cases, it lasts up to three days. Other side effects include farsightedness, mouth, locomotor, confusion and memory disturbances. In the case where people overdose it could cause coma, delirium and respiratory paralysis in severe cases it causes death. Relatively low dosages of the plant have minimal side effects.

Homeopathic

Homeopathic treatments are used in homeopathy. It was developed by Samuel Hahnemann. It is was developed on a basis of a theory called treating like with like. The manner in which homeopathic remedies function is not well defined. Therefore grounds for the safety of the treatment is not well established. For the few cases in which homeopathic treatment was used in the treatment of fibromyalgia notable progress was noted. This led to further trials to test whether the treatment can be used to treat rheumatoid arthritis and osteoarthritis. The results gotten so far are inconsistent. Under the theory of treating a like with a like, it was noted that symptoms noted among individuals suffering from a particular disease are similar to those noted in a healthy person who was already treated for the same disease

(Kassam et al., 2010). The remedies are developed by watering down active substances known to produce similar symptoms. This is done with the belief that it will reduce the harm caused by the illness. This form of treatment requires high skills. For this reason, qualified homeopaths are allowed to prescribe treatment based on the diagnosis. Treatment is done through the administration of drops, homeopathic granules, powders and tablets. Most of these forms of medication is readily available in pharmaceutical shops. Homeopathic treatment bears many similarities to traditional Chinese medicine. The most common challenge of using homeopathic remedies is that there are instances where the remedies are extremely diluted to a point where there are no present molecules and hence no active ingredients available.

Hydroxycitric Acid

Hydroxycitric acid is a chemical. It is said to bear similarities to citric acid. Both acids are obtained from fruits more so those from the Garcinia species. It can also be obtained from hibiscus flowers. Hydroxycitric acid is well known to facilitate weight loss and also improve athletic performance. With regards to weight loss, Hydroxycitric acid is said to inhibit fat storage and also a lower appetite for food. For athletes, Hydroxycitric acid affects exercise performance by controlling the extent in which energy stored in the muscles is used. It also helps reduce fatigue. Untrained women and athletes increase their exercise ability. The acid is consumed for 5 days. However, there is not much research to expand on this knowledge. There is not much information on whether Hydroxycitric acid affects athletes on a long-term basis. With regards to weight loss, Hydroxycitric acid helps those dealing with obesity. Studies show that by consuming Hydroxycitric acid for 8 weeks people improve their weight loss. However if consumed for two weeks not much change is noted. It has also been used to deal with diabetes and high cholesterol among other conditions. Hydroxycitric acid is administered by mouth and is considered safe for consumption. This is in condition that it is consumed for a period of 12 weeks or less. Some of the side effects include headache, nausea or stomach discomfort. Long-term side effects of Hydroxycitric acid have not been established.

Dosage of Hydroxycitric acid varies according to a number of factors among them user's health, presence of underlying conditions and age among others. Predetermined appropriate dosage for Hydroxycitric acid has not been established. As it is a natural product, it is important for users to seek physician's advice.

Hydrogen Cyanide

Hydrogen cyanide is a chemical substance bearing the chemical formula of HCN. It is also called prussic acid. The liquid is colorless, flammable and highly poisonous. Its boiling point is slightly above 78 Farads. The liquid is produced in both large-scale and small-scale. During large-scale production, the liquid is used in the production of adiponitrile and potassium cyanide which are then used in the production of plastics and mining. In nature, Hydrogen cyanide is made up of nitrogen and carbon. The bond between hydrogen and carbon. The acid has a pH of 9.2 and is therefore considered to be weakly acidic (Jiang et al., 2019). When added to water, hydrogen cyanide becomes partially ionized resulting in the production of cyanide anion. When water is mixed with hydrogen cyanide, it results in the production of hydrocyanic acid. Cyanide anion can create salts known as cyanides. Hydrogen cyanide is not easily detectable. This is because it possesses a genetic trait that is recessive. Its odor is a bitter almond-like smell. The liquid used in the killing of whales, rodenticide and poison.

It was also noted that stimulation of muscarinic cholinergic receptors results in the production of hydrogen cyanide. Those receptors are noted in cultured pheo-chromocytoma cells. However, in a living organism, the stimulation is said to lower the production of hydrogen cyanide. Hydrogen cyanide has medical advantages as it kills pathogens such as fungi and bacteria. Destruction of the pathogens because of the toxic nature of hydrogen cyanide.

Hydroxymethylglutarate

Hydroxymethylglutarate is a component utilized by cells in the body in the making of CoQ10. It is considered a cholesterol metabolic precursor. The CoQ10 generated in the cell is an essential nutrient that allows the body to utilize oxygen in the generation of significantly large amounts of energy. Hydroxymethylglutarate is usually given to individuals whose bodies cannot generate sufficient amounts of energy to meet the demand for the entire body. This problem is noted among a significant number of individuals. To check the body's ability to utilize coenzyme Q10, markers such as Malate, Lactate, Fumarate and Succinate are checked. Restrained production of CoQ10 can be noted by high levels of Hydroxymethylglutarate. The elevated levels of Hydroxymethylglutarate may be due to the lowering of cholesterol statin drugs. Elevated levels of Hydroxymethylglutarate may require patients to take CoQ10 supplements.

Hormone Replacement Therapy

Hormone replacement therapy (HRT) is also referred to as menopausal hormone therapy. This form of treatment is mostly used among women to help in the balancing of progesterone and estrogen levels more so during menopause. Other conditions may prompt doctors to prescribe supplementary sex hormones. The treatment is highly effective in treating hot flashes and sweating among other symptoms of menopause. It also minimizes the risk of osteoporosis. There are different types of HRT. One type contains both estrogen and progesterone while others only have estrogen or testosterone. Depending on the diagnosis, the doctor can come up with a dosage having the right type and amount of hormones. Among the areas of application of HRT is menopause (Jima & Megersa, 2018). Although it is not an illness, some women as unable to deal with some of the symptoms of menopause. Some of the symptoms of menopause include osteoporosis, night sweats, sleep problems, hot flashes, urinary problems, irregular periods, vaginal dryness, reduced concentration, mood changes and thinning hair.

Figure 39: HRT controls production of hormones.

Source: https://www.google.com/url?sa=i&url=https%3A%2F%2Fwww.openaccessgovernment.org%2Fhormone-replacement-therapy-boost-your-immune-system%2F85516%2F&psig=AOvVaw3vSHXGlsbTkyEjGN1Fcl7U&ust=1633679294510000&source=images&cd=vfe-&ved=0CAsQjRxqFwoTCPDQrOnnt_MCFQAAAAAdAAAAABAM

Other than the symptoms of menopause, HRT can be used to improve muscle function, minimize skin aging, mortality among individuals who suffer from post-menopause and reduce risks of heart attacks and heart failure. In other cases, hormonal treatment is used in treating prostate cancer, low blood testosterone and birth control.

Although HRT is useful in dealing with symptoms of menopause, there are several side effects noted among patients. They include leg cramps, acne, breast tenderness, headaches, indigestion, migraines, bloating, depression, and back pain, mood changes, swelling in different parts of the body, vaginal bleeding and nausea. It was noted that most of the side effects disappear after a week or two. Despite the numerous benefits of the treatment, it does not apply to everyone. For this reason, there are certain factors taken to consideration before the treatment is administered. The treatment is not administered if the patient had suffered from gallbladder disease, high blood pressure, heart disease, uncontrolled hypertension, stroke, blood clots, breast cancer and elevated levels of triglycerides in the blood.

I

Indian Bael

Indian bael is a plant whose branches, fruits, leaves and roots have medicinal value. It was traditionally used in the treatment of diabetes, constipation and diarrhea among others. Modern research aims at proving the medicinal uses. The medicinal value of the plant is attributed to its chemical composition of the plant. Among the chemicals found in Indian bael include coumarins, tannins and flavonoids. The chemicals are known to help minimize inflammation. The chemicals are also useful in the treatment of diarrhea and asthma among other illnesses. The chemicals are also utilized in lowering blood sugar levels.

Synonym

Other names used to refer Indian bael include vilvam, Aegle marmelos, Manzana de Piedra, Bael tree, shivaduma, bilva, Arbre de Bael, Pomme du Bengale, Bel Indien, Stone apple and Bengal quince.

Usage

Indian bael has been used in making several medicinal products used in treating a variety of diseases. Among the diseases treated using Indian bael include asthma. Products having components of Indian bael are said to help people suffering from asthma breathe better. Usually, the Indian bael fruit is combined with Boswellia for the treatment to be effective (Helmstädter & Staiger, 2014). Other than asthma, Indian bael is used in treating diabetes, diarrhea, stomachache, constipation, snakebites and depression. It is also effective in improving memory. Dried Indian bael fruit powder is administered to patients to deal with a bacterial infection called shigellosis.

Dosage

There are a number of factors that influence the kind of dosage given to patients. They include underlying conditions, patient's age and health. The appropriate dosage of Indian bael has not been scientifically determined. For this reason, patients are advised to seek and follow consults from their pharmacists. Also, medicine containing Indian bael should be consumed according to the prescription.

Possible side effects

As a natural medicine, Indian bael is safe for consumption. This is when the dosage is taken in small amounts. When ingested orally, large amounts of Indian bael is said to cause constipation and an upset stomach. There is little information to indicate some of the side effects of using Indian bael on the skin.

Extracts

Ipomea Aquatica

Ipomea aquatic is a plant cultivated as a vegetable. It is also called water convolvus, water spinach, river spinach or water morning glory. The plant produces tender shoots and is commonly grown in Asia. It is grown for food and was used traditionally in medicine. It was used in treating high blood pressure, nosebleeds and piles. Modern medicine shows that the plant could be used in cancer treatment as it limits the growth of cancer in the body. It, therefore, contains anti-cancer activity. The challenge of using Ipomea aquatic in the treatment process is that when it is not properly utilized it could cause some infections.

Immortalitea

Immortalitea commonly known as Green chai is a plant used to make green tea. It is also used to make green tea products containing other elements such as rose petals, rosemary, lemons and sage. Immortalitea is blended with other plants creating a herbal mixture (Gupta et al., 2018). Brewed green tea made using Immortalitea contains antibacterial, anti-aging, antioxidant and antiviral properties. The made cup of tea is said to be soothing and can therefore be used to reduce stress or fatigue. Immortalitea is said to be dull compared to tea made traditionally.

Iodine

Iodine is a chemical component heavily needed by the body. However, the body does not produce it. The major source of iodine in the body is food. A little amount of iodine in food is needed. Naturally, iodine is obtained from the ocean which has high concentrations brought about by seaweed. Hormones are produced by the thyroid glands. One component needed in the production is iodine. Insufficient amounts of iodine in the body causes the thyroid gland to work much harder. An overworked thyroid gland enlarges evidenced by a swollen neck. This condition is known as goiter. Very low amounts of iodine in the body could cause severe implications. As thyroid glands are concerned with hormone production, low iodine levels could cause infertility in women as an absence of some hormones cause women to stop ovulating. The deficiency of iodine could result in thyroid cancer. This is as a result of autoimmune disease. Studies show that iodine deficiency increase the risk of other cancers. Iodine deficiency among pregnant women could result in mental problems for the baby and the mother could suffer from high blood pressure.

Iodine is used in preventing thyroid disorders, goiter, mastalgia and fibrocystic breast disease. Other uses of iodine include it being used in radiation procedures in a bid to hinder radioactive iodides from reaching thyroid glands. Tablets of potassium iodide is readily available to the public as food supplements. In cases of skin inflammation, iodine is applied to the skin. It kills germs and creates a conducive environment that would allows healing. It is also used to eradicate soreness in the digestive tracts and the mouth. For women who have undergone a cesarean, iodine is used to prevent the risk of swelling in the uterus walls.

Iron

Iron is a vital mineral found in myoglobin and hemoglobin. In red blood cells, iron is needed to allow the transportation of carbon dioxide and oxygen. Foods rich in iron include cereals, meat, spinach, fish and tofu. Low levels of iron in the body cause anemia which in severe cases causes complications in the kidney and during pregnancy. Consumption of iodine helps prevent anemia.

Indian Vanilla Beans

Indian vanilla bean is a plant that produces vanilla beans. The beans are enclosed within a pod. The beans have both medicinal and flavoring properties. However, purchasing of Indian vanilla beans can be quite expensive resulting in the production of laboratory-produced vanilla as a substitute. In other cases, extracts are used in diluting Indian vanilla beans (Griffiths et al., 2002). However, certain extracts have been banned by the FDA as it contains certain chemicals. Some of the medicinal uses of Indian vanilla beans is that it is used in treating tooth decay, lowering anxiety, intestinal gas and increased sexual desire. It is also used in creating syrups used to flavor medication.

Irish Breakfast Tea

Irish breakfast tea is tea made by blending a variety of tea brands. Some of the tea brands used include Ceylon and Assam teas. Irish breakfast tea is commonly found in Ireland. Irish breakfast tea is served with milk. This is because the Irish breakfast tea is said to be very strong though there are cases where people drink black tea. Honey or sugar is used to sweeten the tea. Irish breakfast tea when compared to other brands of tea has a higher caffeine composition and has a very robust taste. Irish breakfast tea is red in color.

Italian Herb Seasoning

Italian herb seasoning is a combination of herbs blended and is usually used in making a number of Italian dishes. Some of the herbs used in making Italian herb seasoning include marjoram, oregano and basil. Most of the herbs are readily available in Italy. The spaces made their way to America through immigrants who wanted to ensure their food was filled with flavor. The Italian herb seasoning had all the seasoning needed by individuals. It is considered to be a single step to adding flavor to food. Italian herb seasoning has a number of components that have medicinal value. Seasoning has been used in treating gastrointestinal illness as it possesses the ability to protect the body from food-borne illness. Seasoning possesses the ability to kill bacteria. It possesses antimicrobial properties.

It is also useful in digestion. Stimulation of bile juice is due to oregano. It also helps deal with cases of bloating. Herbs such as thyme possess oil that helps reduce bloating. Italian herb seasoning is also useful in cancer prevention. Antioxidant properties found in thymol is used to reduce risk of cancer. The seasoning contains herbs which is useful in preventing constipation as it makes the flow of food in the gut easier.

Infused Oils

Many remedies are used in making infused oils. Due to the medicinal properties of herbs used in making infused oils, they can be used to treat a variety of illnesses. Usually, infused oils are used to treat skin diseases. In other cases, they are ingested. Infused oils also have therapeutic advantages (Fitzgerald et al., 2020). For children suffering from diaper rash, infused oil having calendula is used. The same infused oil is used in treating psoriasis and eczema. Infused oil containing Hypericum perforatum is used to treat sunburns. Infused oils are used based on the doctor's prescription.

Inhibitory Concentration

Inhibitory concentration are used in defining CMV isolate resistance. The various ranges of concentrations are related to different mechanisms. Concentration ranges larger than 1.5-3 micrograms per milligram is linked to 2 mechanisms. They include polymerase related to point mutations occurring in the viral DNA and lower intracellular phosphorylation. Two activities contribute to lower intracellular phosphorylation including deletion in phosphotransferase or point mutations.

Immunoglobulin-E igs

Among the tests done on blood is the immunoglobulin-E test. The test evaluates the amounts of IgE antibodies in the blood. Immunoglobulins are also known as antibodies. They are a vital element found in blood that helps to ensure the body is protected from pathogens that cause diseases. Antibodies can identify germs and kill them. In as much as immunoglobulin-E is needed in the body, it should be at a relatively low amount. In special cases where the amount of immunoglobulin-E is in high amounts then it is likely that the body tends to overreact in the presence of allergens. There are different kinds of immunoglobulin-E based on the allergen they react to. In the case of immunoglobulin-E igs, it reacts to allergen igs.

Intragastric

The term intragastric is means occur in the stomach. It is a form of medical administration. Intragastric balloon therapy is used in treating obesity. There are different kinds of intragastric balloons used in the treatment process. Spatz, ReShape, Obera and Obalon is used in weight loss while ReShape and Obera are used to normalize metabolic parameters. There are a number of risks associated with intragastric administration of drugs more so with balloon therapy.

Inhalation

Inhalation is the process in which the body draws air into the body. Other than air, the body could inhale dust, mist, vapor and fumes. Inhale substances find their way into the lungs and finally the bloods stream. Inhalation is also a method used traditionally and currently in administering herbal medicine. Inhaled herbal medicines are extensively used in treating chronic diseases like asthma. Some of the inhaled herbal medicines are effective in opening up breathing passages.

Interleukin

As humans communicate with one another to pass vital information, in the same manner do cells in our bodies communicate. Communication between cells is mediated by proteins generated by the body known as Interleukin. Other than facilitating communication, these proteins also facilitate motility and control cell growth. They are also involved in the stimulation of many activities among them immune responses (Eddouks et al., 2017). This group of proteins along with other molecular messengers can be grouped under a group called cytokines. These molecular messengers are actively involved in controlling cell behavior. Production of Interleukins requires the presence of a stimulus. This means that these groups of proteins are not stored in cells. When produced, these proteins find their way to the affected cell. Binding occurs on the cell surface and is facilitated by receptors. The result of the binding process causes signals to be transmitted in the cell and modifies its behavior.

Intramuscular

This form of treatment involves drugs being directly administered into a patient's muscles. In most cases, the drug is injected into the identified or specified region. In most cases, the drug is administered to the center of the muscles.

Intradermal

Intradermal is a technique used in administering treatments to patients. It is commonly denoted as ID. This involved a superficial or shallow injection of the medicinal substance into the skin. The skin has many layers. The treatment is administered between the hypodermis and epidermis. This technique of medication administration is seldom used compared to injecting medication into the subcutaneous muscle or tissue. This technique of medication administration is used in specific cases. It is not preferred for the normal administration of drugs. ID is used during specific therapies to conduct tests. The advantage of this technique of administration is that it improves immune responses when vaccinations are administered, increases drug uptake and extensive use in cancer treatment. Intradermal injection is mostly used when small molecules and soluble proteins are involved. It improves the rate of drug uptake thereby allowing medication to reach targeted sites at a faster rate. This technique also makes it easy to note the body's reaction to the administered drug. An important tool used in this technique is a syringe. Calibrations may vary depending on the amount of dosage to be administered. The needles used should be strong allowing entry into the body.

Infusion

Infusion is a treatment technique used in infusion therapy. This technique involves medication being administered to patients through a catheter. It is the alternative for of medicine administration when it cannot be consumed orally. It is also used when the amount of dosage to be administered is to be strongly controlled. This form of treatment is said to be intravenous. There are different types of infusion therapy which include subcutaneous, intramuscular and epidural. This form of treatment is highly effective if the drug to be administered can lose its effectiveness when it enters the gut. IVs are administered when the patient needs to stay hydrated. It is also used when quick drugs need to reach targeted body parts faster (Dhama et al., 2018). An example of infusion therapy is when insulin has to be administered under the skin through an insulin pump. This form of treatment is also used when the patient is unconscious. It helps delivers nutritional fluids. Some of the medications delivered through infusion therapy include growth hormones, inotropic heart medication, antibiotics, antivirals, immunotherapy, antifungals, blood factors, immunoglobulin replacement, biologics and chemotherapy.

Intraperitoneal

This is a technique of drug administration in which drugs enter the body through the peritoneum. Abdominal organs like the intestines are enclosed by a membrane. The membrane is transparent and thin. The same membrane also protects the walls of abdominal cavities. The peritoneum performs a number of functions among them performing peritoneal dialysis among patients suffering from renal failure. Its cavity has physiologic traits that aid in the removal of toxic metabolites and allow the administration of medication into the body. The intraperitoneal route is used to administer antibiotics which help in the treatment of peritonitis. It has also gained popularity among doctors for chemotherapy for individuals suffering from gastrointestinal and gynecological cancer. There is a lot of care taken with the administration of drugs. Some of the factors involved in the administration of drugs include surface area, time of contact, peritoneal inflammation and peritoneal blood flow.

Intrathecal

Intrathecal is a technique used to administer medication to patients. This technique involves medication being administered to the cerebrospinal fluid through the spinal cord. The technique is used to control pain, and administer spinal anesthesia and chemotherapy. This method of administration is mostly used when the drug has to be administered to certain areas within the body and helps medicine get to areas of need. Therefore, this method of treatment helps prevent bacterial infection more so after surgery. Drugs administered through this method are those that should not be hindered by blood-brain hindrance. Medications administered through this method are specially made by a pharmacist as there are certain elements found in normal medicine that cannot be administered directly to the spinal cord or to the brain (Devkota & Watanabe, 2020). The treatment works by having an intrathecal pump implanted under the skin or region where the drug is to be administered. A tube known as a catheter is used to create the link between the pump and the area where the drug is to be administered. If the drug is to be administered to the spine then the catheter is placed at the base of the spine. The drug is pumped into the spine. The dosage administered through this method is relatively smaller than the amount consumed through the mouth. A common side effect of administering drugs through this method is sleepiness. Another challenge if this method is that there high risks of an infection occurring.

Intravenous

Among the methods of administering medication is through intravenous infusion or injection. This method of administration involves medication being channeled directly to the vein. The tool used in this case is a tube. As medication is delivered to the vein, the method got the name intravenous. The method works by having a needle or a catheter inserted to a particular vein. In most cases, the tube stays on the vein for as long as medication has to be administered. It is removed at a given frequency to reduce the risk of infection. As the tube is inserted, the number of needle injections reduces as doses will be administered through the tube. This method of drug administration cannot be self-done. A professional may be involved in the process.

One of the biggest advantages of this method of drug administration is that it allows pharmacists to control the amount of dosage administered to patients. It is also administered in emergencies where the drug must get to the targeted are quick. Poisoning, stroke and heart attacks are examples of the emergency situations. This form of medicine administration allows medication to get to the blood stream fast. Though this form of administration allows medicine get to the targeted area directly, proper care should be taken so that the right amount of dosage is administered as per the pharmacist's prescription.

J

Jiaogulan

Synonyms

Adaptogen, Adaptogène, Amachazuru, Dungkulcha, Fairy Herb, Ginseng du Sud, Gynostemma, Gynostemma pedatum, Gynostemma pentaphyllum, Herbe de l'Immortalité, Immortality Herb, Jiao Chu Lan, Jiao Gu Lan, Miracle Grass, Penta Tea, Plante de l'Immortalité, Southern Ginseng, Thé de l'Immortalité, Vitis pentaphylla, Xian Cao, Xianxao.

Usage

Jiaogulan is a wild plant found in China. The leaf is employed in the production of medication. Jiaogulan is also known as "Southern Ginseng" since it grows in south-central China and is utilized similarly to ginseng.

Jiaogulan is a Chinese herb that is used to treat high cholesterol, and high blood pressure, and to improve heart function. It is also used to boost the immune system, increase stamina and endurance, boost resistance to environmental stress (as an "adaptogen"), improve memory, and prevent hair loss.

Treatment of low appetite, cough, chronic bronchitis, persistent stomach pain (chronic gastritis), pain and inflammation, ulcers, constipation, stress, gallstones, obesity, cancer, diabetes, insomnia, backache, and pain are some of the other uses(Miara et al., 2018).

According to preliminary studies, consuming jiaogulan three times daily by mouth for four months does not enhance markers of liver function, body mass index, cholesterol levels, kidney function, or blood sugar in persons with nonalcoholic fatty liver disease.

Jiaogulan is used by some as an anti-aging, antioxidant, and detoxifying agent.

Dosage

For high cholesterol: jiaogulan extract 10 mg three times daily.

For diabetes: 3 grams of jiaogulan tea in 60 mL of water has been taken twice daily for up to 12 weeks.

Side Effects

Jiaogulan is safe when taken by mouth short-term (up to 4 months). It can cause some side effects such as severe nausea and increased bowel movements.

Extracts

Jiaogulan is a herb used in traditional Chinese medicine. During the Ming Dynasty, Zhu Xiao first described it in his book "Herbs against hunger" in 1406.

Xiancao, the second Chinese name, means "herb of immortality." People in South China drink Jiaogulan tea regularly, and the number of people living to be over 100 years old in the Guizhou region is above average (Nadda et al., 2020).

Jiaogulan is a kind of adaptogen. It means that the plant can adapt to the environment and has a direct impact on the body's weak points. This also implies that Jiaogulan may have varying effects on various persons. Furthermore, adaptogenic implies that the plant does not introduce a new component, such as negative effects.

Jiaogulan's essential substances are significant.

The high amount of gypenosides, which are saponins, is a unique feature of this plant. It also contains ginsenosides, which are the same chemicals found in ginseng. Jiaogulan has more than 80 saponins compared to Panax ginseng's 28. Polysaccharides, amino acids, vitamins, and minerals are among the other constituents.

Fairvital Jiaogulan capsules are an adaptogenic natural antioxidant with a 4-fold concentrated extract corresponding to 2000mg Jiaogulan in each capsule. They are vegetarian and vegan, gluten-free, lactose-free, and fructose-free, and devoid of magnesium stearate and silicon dioxide (Thakur, 2011)..

Juniper

Synonyms

Baie de Genévrier, Common Juniper, Common Juniper Berry, Enebro, Extract of Juniper, Extrait de Genévrier, Genévrier, Genévrier Commun, Genievre, Genièvre, Ginepro, Huile de Baies de Genévrier, Huile de Genévrier, Juniper Berry, Juniper Berry Oil, Juniper Extract, Juniper Oil, Juniperi Fructus, Juniperus communis, Oil of Juniper, Wacholderbeeren, Zimbro.

Uses

Juniper is a short to medium-sized tree that grows naturally in portions of Europe, North America, and Asia. There are numerous juniper species, but Juniperus communis is the most common in North America.

The juniper berry is used to create medication. Juniper berry extract and juniper berry essential oil are both used in medicinal formulations. Juniper berry oil should not be confused with cade oil, which is made from juniper wood (Juniperus oxycedrus) (Stanisavljević et al., 2007).

Juniper is used to treat digestive issues such as upset stomach, intestinal gas (flatulence), heartburn, bloating, and loss of appetite, as well as GI infections and intestinal worms. It is also used to treat urinary tract infections (UTIs) as well as kidney and bladder stones. Other applications include snakebite treatment, diabetic treatment, and cancer treatment.

Some individuals apply juniper directly to the skin to treat wounds and joint and muscular discomfort. Juniper essential oil is used to cure bronchitis and numb discomfort.

The juniper berry is commonly used in dishes as a condiment and as a flavoring element in gin and bitter concoctions. The extract and essential oil are used to flavor foods.

Juniper oil is used in the production of soaps and cosmetics as a scent.

Figure 40: Juniper Berries.

Source: https://www.healthline.com/nutrition/juniper-berries

Juniper extract and oil are used in a variety of cosmetics, including lipstick, foundation, hair conditioners, bath oils, bubble baths, eye shadows, and many others (Seth, 2003).

Juniper berries include compounds that may reduce inflammation and gas production. It may also be beneficial against germs and viruses. Juniper may also make you feel the need to urinate more frequently.

It is generally used for:

- Stomach ache.
- Heartburn.
- Bloating.
- Appetite loss.
- Infections of the urinary tract (UTIs).
- Stones in the kidneys and bladder

- Muscle and joint pain
- Wounds.
- Other circumstances.

Dosage

Generally, 2 to 10 g/day of the whole, crushed, or powdered fruit (corresponding to 20 to 100 mg of essential oil) has been used for dyspepsia.

Essential oil: 0.02 to 0.1 mL 3 times daily.

Fluid extract: 1:1 (g/mL); 2 to 3 mL 3 times daily.

Infusion: 2 to 3 g steeped in 150 mL of boiled water for 20 minutes 3 times daily.

The optimal dose of juniper is determined by numerous factors, including the user's age, health, and other conditions. There is currently insufficient scientific evidence to identify a suitable range of juniper dosages. Keep in mind that natural products are not always safe, and dosages can be critical. Before using, be sure to read the product label and consult your pharmacist, physician, or another healthcare expert. Large doses of juniper berries may cause catharsis and convulsions. The juniper volatile oil may be nephrotoxic (Ningthoujam et al., 2012).

Side Effects

When ingested in typical food proportions, juniper, juniper berry, and juniper extract are all safe.

Juniper is safe for most individuals when taken by mouth in tiny doses for a short period, inhaled as a vapor, or applied to the skin in limited areas. When applied to the skin, juniper can cause irritation, burning, redness, and swelling. It should not be used on big skin wounds.

Taking juniper by mouth for an extended period of time or in excessive doses is dangerous because it can induce kidney issues, seizures, and other significant side effects.

Extracts

CEP (part of the Solabia Group) manufactures a wide range of plant extracts using a variety of extraction methods.

Extract of Juniper Berry is available in the following formats: Vegebios 1.5P

Vegebios 1.5P - Steam distillation that gives a water soluble herbal distillate. Preserved with Phenoxyethanol.

Jimbu

Figure 41: Jimbu.

Source: https://myrepublica.nagariknetwork.com/mycity/news/health-benefits-of-jimbu

Synonyms

Jimbu, jambu, jamboo, or jhiku-cha

Jambu is an onion family herb that is widely utilized in Nepal and some central Himalayan states of India, such as Uttarakhand, where it is known as Jamboo / Faran. It is made up of two Allium species, *A. hypsistum and A. przewalskianum*. The herb, which has a flavor that is similar to onion and chives, is most usually used dried. It is used to flavor vegetables, pickles, and meat in Nepal's Mustang region. It is most typically used to flavor urad dal or lentils in the remainder of Nepal (Saslis-Lagoudakis et al., 2014). To bring forth the flavor of the dried leaves, they are fried in ghee. Because Jimbu is a seasonal plant, people dry it and store it for later use (main harvest between June and September)

Uses

The herb, which has a flavor that is similar to onion and chives, is most usually used dried. Jimbu, which has traditionally been utilized as a major cooking spice, provides the following health benefits.

It is used to treat a variety of medical conditions, including flu and high altitude sickness.

It is used to treat coughs and colds, and it can heal sore throats.

It also aids in the treatment of gastric disorders.

Jimbu is high in vitamin C and aids in the inhibition of the formation of dangerous bacteria in the body.

Jimbu also aids in the development of red blood cells and the maintenance of healthy skin.

Dosage

The optimal dose of jimbu is determined by numerous factors, including the user's age, health, and other conditions. There is currently insufficient scientific evidence to identify a suitable range of juniper dosages. Keep in mind that natural products are not always safe, and dosages can be critical. Before using, be sure to read the product label and consult your pharmacist, physician, or other healthcare experts (Nazim et al., 2020).

Side Effects

No side effects known

Extracts

Jimbu is ordinarily used as a dried herb

Jakhya

Synonyms

Jakhya (Hindi: ; Urdu:) (also known as dog mustard or wild mustard) is the seed of the Cleome viscosa plant that is used to season food. It is primarily farmed and consumed in Uttarakhand, as well as in India's and Nepal's Terai areas.

When roasted in oil, the seeds get a dark brown color and crackle. It is popular in Garhwali and Kumaoni cuisines.

Jakhya Scientific Name: Cleome viscosa

Uses and Extracts

Jakhiya grows in the tropics all over the world and is used in traditional medicine in many parts of India and beyond. Almost every part of the plant is utilized to treat ailments.

According to the Council for Scientific and Industrial Research's Treatise on Indian Medicinal Plants, the plant is used to cure fever, inflammations, liver disorders, bronchitis, and diarrhea (Tuttolomondo et al., 2014). The oil derived from jakhiya seeds is therapeutic. Crushed seed oil is used to treat infantile convulsions and mental problems. Several published research have demonstrated that Cleome viscosa leaves can help cure wounds and ulcers.

Dosage

No scientific evidence for therapeutic dosage. Mostly used in cooking.

Side Effects

Common side-effects: Not known.

Not so common side-effects: Not known.

Pregnant or Nursing Mothers: Not known.

Children: Not known.

Jalapeno

Synonyms

The jalapeno is a medium-sized cultivar of the chile pepper pod species *Capsicum annuum*. A ripe jalapeo chile measures 5–10 cm (2–4 in) in length and has a round, firm, smooth flesh that measures 25–38 mm (1–11/12 in) in width. It has a pungency range of 3,500 to 8,000 Scoville heat units. Although it is typically harvested and consumed when still green, it is occasionally allowed to fully mature and turn red, orange, or yellow. It is broader and milder than the comparable Serrano pepper (Teixidor-Toneu et al., 2018).

It goes by several vernacular names such as Lalmirchi in Hindi, Lanka Maricha in Bengali, Menasinakai in Kannada, Molagu in Malayalam, Lal mircho in Marathi, Mircha Kaya in Telugu, and Milagai in Tamil.

Uses

Jalapenos are low in calories and high in nutrients such as vitamins, minerals, fiber, and antioxidants. Being high in vitamins A and C, which act as powerful antioxidants, helps counteract free radical damage to cells, boosts the immune system, and maintains the skin tight and beautiful. It is also high in vitamins B6 and folate, which are involved in a variety of biological activities. Aside from this, one of the most active components in jalapenos is capsaicin, an alkaloid that gives peppers their characteristic spicy flavor and has a plethora of health benefits.

Jalapenos can help you lose weight by increasing your metabolism, burning fat, and decreasing your hunger. The vast array of active substances found in jalapeno, including capsaicin and capsaicinoids, has been shown to increase metabolism by 5%, making it simpler to shed those stubborn fats. Fresh jalapeño is high in vitamin C, providing about 118 mcg or over 200 percent of the recommended amount per serving. Vitamin C, being a powerful antioxidant, is required for collagen formation in the system, which is essential for maintaining the integrity of blood vessels, skin, tissues, organs, and bones (Silva et al., 2010). Capsaicin, the main component in jalapenos, is recognized for its tremendous anti-cancer properties, which are exceptional for eliminating over 40 forms of cancer without affecting normal cells. Stomach ulcers are a common health disease that affects the majority of individuals and are caused by the growth of the H. Pylori bacteria, increased acid secretion, poor blood flow, stress, excessive alcohol intake, and smoking. Spicy foods, such as jalapenos, are thought to aggravate stomach ulcers; however, investigations have shown that this is not the case.

Figure 42: Cinder Jalapeno Pepper.

Source: https://hosstools.com/product/cinder-jalapeno-pepper/

An annual dose of capsaicin is thought to reduce bad cholesterol levels and platelet aggregation. Jalapeno, which is high in vitamin C, A, and bioflavonoids, improves blood circulation and dilates blood vessels. Jalapenos include a variety of bioactive chemicals and vitamins that can inhibit the growth of foodborne bacteria and yeast. Jalapeno extracts can prevent cholera bacteria from generating toxins, reducing the severity of this foodborne sickness.

Dosage

Capsaicin food sources may not contain enough of this component to give all of its potential benefits. As a result, many take pills with a higher dosage of capsaicin.

Although dosages vary, several studies employ 2–6 mg of capsaicin each day.

Because cayenne pepper is high in capsaicin, it is used as the principal active ingredient in the majority of supplements on the market. Cayenne pepper capsules typically include 500 mg of cayenne pepper in each capsule.

Side Effects

A burning sensation throughout your digestive tract is one of the most prevalent side effects of capsaicin pills.

Capsaicin-containing meals or supplements may aggravate acid reflux or heartburn symptoms.

Your digestive tract grows desensitized to capsaicin's effects over time, decreasing any discomfort you may feel.

To avoid this side effect, take the dose right before a meal, as food acts as a buffer to reduce the burning sensation.

Extracts

Capsaicin is the main compound extracted from jalapenos and it also happens to be the compound that gives plants in the capsicum family their spicy hot notes. It also happens to be a compound with medicinal uses. Extracts can be homemade by grinding, boiling and reducing the jalapenos (Peng et al., 2017).

Juniperus Communis

Synonym

Juniperus communis, sometimes known as common juniper, is a tiny tree or shrub in the cypress family Cupressaceae. This evergreen conifer has the broadest geographical distribution of any woody plant, spanning the mild temperate Northern Hemisphere from the Arctic south in mountains to roughly 30°N latitude in North America, Europe, and Asia. Relict populations can be found in Africa's Atlas Mountains.

Don't confuse juniper berry oil with cade oil, which is distilled from juniper wood (*Juniperus oxycedrus*).

Uses

- Bloating.
- Heartburn.
- Joint and muscle pain.
- Kidney and bladder stones.
- Loss of appetite.
- Upset stomach
- Improves digestions
- Improves sleep
- Urinary tract infections (UTIs).
- Wounds.
- Juniperus communis
- Natural antiseptic
- Other conditions.

Side Effects

While juniper berries are typically healthy for the majority of people, there are some warnings and therapeutic interactions to be aware of.

To begin, pregnant women should never consume juniper berries, either whole or in essential oil form, as they may cause harm to the unborn child or force uterine contractions. Juniper is also not advised for people who have impaired kidney function (Sepúlveda-Cuellar et al., 2021).

An allergic reaction to juniper berries is possible, which could appear as skin problems (such as a rash) or respiratory problems. If you develop any of these symptoms after using juniper berries, stop using them immediately and visit your doctor.

Dosage and Extracts

Capsules: Follow Mfr's instructions (eg, 800 mg PO BID)

Berries: 1-2 g of berry PO TID

Tea : One cup PO TID Steep 1 tsp crushed berry (2-3 g) in 150 mL boiling water x 10 min.

Essential oil: 20-100 mg; no more than 4 weeks

Liquid Extract (1:1 in 25% alcohol) :2-4 mL PO TID

Tincture (1:5 in 45% alcohol): 1-2 mL PO TID

Berry oil (1:5 in 45% alcohol): 0.03-0.2 mL PO TID

K

Katuk

Synonyms

Star Gooseberry Sweet Leaf Sweet Leaf Bush

Katuk (Botanical name *Sauropus androgynous*) is a shrub-like plant of the Euphorbiaceae family. This herb is typically seen growing in high-temperature, humid environments. It is a native Southeast Asian vegetable that is widely cultivated for use in traditional medicine in this region of the world. This herb grows upright with several branches. It typically grows to a height of 2.5 meters and has oval-shaped deep green leaves. The katuk plant has the appearance of a clump (Salim et al., 2019). This herb's branches are fairly delicate and contain divided leaves that emerge alternately on the stem. The leaves are around 2.5 cm long and 1.25 cm to 3 cm wide. Katuk flowers can be found alone or in groups of three. The little fruit, which measures around 1.25 cm across, is carried by a long stem. Katuk is one of Southeast and South Asia's most popular leafy vegetables. Furthermore, this plant is notable for its palatability as well as its extremely high yields.

Uses

Sauropus androgynus, often known as katuk, has been discovered to be a therapeutic herb with high antioxidant potential. Katuk leaves have traditionally been used to cure various ailments and to aid in weight loss. They are also eaten as a vegetable, and the leaves of this plant are used to make a variety of cuisines. Katuk leaves contain an adequate amount of several macronutrients as well as virtually all of the micronutrients. Carotenoids, phenolic compounds, several antioxidant vitamins, and numerous important minerals are among the micronutrients found in katuk leaves.

The leaves of *Sauropus androgynus* are a staple diet in a number of underdeveloped countries, providing the poor population with a variety of critical nutrients. Furthermore, katuk leaves contribute in the maintenance of healthy conditions.

The katuk plant has been widely utilized in traditional remedies from diverse regions of its origin for a number of purposes, including encouraging lactation, treating wounds, easing urinary issues, and acting as an anti-diabetic agent. These leaves have also been shown to be beneficial in the treatment of fever. Aside from their therapeutic uses, the vivid green leaves of katuk are also used as a great culinary coloring additive (Rawat et al., 2020).

The bright green leaves of katuk are a wonderful source of chlorophyll, which is necessary for cell rejuvenation, blood formation, and is also beneficial to the gut flora, circulation system, and proper bowel elimination. Indeed, all plants with high nutritional value, such as katuk, which aids in detoxification, are always a great addition to our daily diet. The leaves of katuk are an excellent natural source of potassium, a vital mineral with numerous health advantages.

An infusion made from katuk leaves is also used as a poultice to cure fevers and repair ulcers. The katuk leaf is also used in folk medicine to treat snoring and teeth grinding while sleeping. As a result, it is recommended that you take katuk leaves regularly in soups and other foods.

People have long referred to the katuk plant as "multi-green," owing to its high quantities of vitamins and other nutrients. The leaves of this herb are commonly consumed raw in salads as well as stir-fried. Curries also make use of the leaves.

Side Effects

Despite the health benefits of katuk leaf, taking too much of it might be hazardous. These leaves contain a considerable amount of papaverine, an alkaloid. Excessive consumption of katuk leaves may cause drowsiness, lightheadedness, and even constipation. When the katuk leaves are taken raw or uncooked, the papaverine present in them may cause vasodilation and potentially result in bronchitis obliterans, a progressive respiratory illness. The majority of the damage is caused by consuming excessive numbers of uncooked leaves as well as drinking the raw juice of these leaves instead of eating boiling or stir-fried katuk leaves.

Dosage

Anything we eat too much of can have a negative health effect. Excessive consumption of raw katuk for weeks on end may not be good for you. It contains a compound named Papaverine that lowers blood pressure and has a string of other medicinal and pharmacological uses. One group of Taiwanese women trying to lose weight by juicing katuk, found out the hard way (Reinaldo et al., 2020). Four middle-aged women, however, who drank 3.5 to 7 ounces (100g to 200g) a day of the extract for up to two months had to have lung transplants within a year and a half of consuming the extract.

Kale

Synonyms

Boerenkool, Borecole, Brassica oleracea, Brassica oleracea var. acephala, Brassica oleracea var. viridis, Chou Fourrager, Chou Kale, Col rizada, Kale Frisé, Kale Frisé Sibérien, Kale Leaf, Kale Noir, Kale Rouge de Russie, Winter Greens.

Kale is a dark, leafy green that can be either raw or cooked.

This superfood has been served on dinner tables since Roman times and has long been popular throughout much of Europe. The vegetable is a member of the cabbage family, which includes broccoli, cauliflower, and collards (Schultz et al., 2020).

Kale is more popular than ever before because it's high in vitamins and minerals.

Kale can be curly or flat, and it can also have a bluish hue mixed in with the green. The flavors are distinct.

Figure 43: Kale.

Source: https://www.healthline.com/nutrition/10-proven-benefits-of-kale

Among them are the following:

Curly kale: the most popular type of kale, with brilliant green ruffled leaves.

Dinosaur kale is distinguished by its tiny, wrinkled green leaves that are linked to a strong stem.

Redbor kale has ruffled leaves that range from bright crimson to purple in hue.

Russian kale is a less common variety with flat leaves with a fringe that ranges from green to scarlet to purple.

Uses

Health Advantages of Kale

Vitamin A (essential for eye and bone health as well as a robust immune system), vitamin C (helps prevent colds and chronic diseases), and vitamin K (good for blood clotting and bone-building)

Folate is a B vitamin that is essential for brain development.

Alpha-linolenic acid is a kind of omega-3 fatty acid. (Although kale contains significantly less omega-3 than fish, it is another method to incorporate this healthful fat into your diet.)

Lutein and zeaxanthin, are minerals found in kale that help prevent macular degeneration and cataracts (Saslis-Lagoudakis et al., 2011).

Phosphorus, potassium, calcium, and zinc are examples of minerals.

Kale has the following nutrients that can benefit you:

Lower your cholesterol

Cancer prevention

Reduce your weight by filling you up with a high water content but few calories

Side Effects

Kale is a terrific complement to almost any diet. However, in high doses, kale and its cabbage family kin can interfere with how your thyroid gland functions.

Ask your doctor about how various foods can impact your thyroid if you have hypothyroidism, often known as an underactive thyroid (Miara et al., 2018).

Kale or cruciferous vegetable allergy can cause a variety of symptoms. These are some examples:

- skin irritations
- hives
- disorientation
- swelling of the lips, tongue, and throat
- digestive discomfort

Kale dosage is determined by numerous factors, including the user's age, health, and other conditions. There is currently insufficient scientific evidence to identify an optimal range of kale dosages. Keep in mind that natural products are not always safe, and dosages can be critical. Before using, be sure to read the product label and consult your pharmacist, physician, or other healthcare experts.

Extracts

Kale Liquid Extract – Standardized is a liquid combination of Kale Extract and Propanediol created using a proprietary manufacturing process.

Through its antioxidant properties, kale, a leafy powerhouse rich in vitamins and carotenoids, protects and strengthens the skin. Kale Extract is high in Vitamin A and Beta-Carotene (a precursor compound of vitamin A). These nutrients are believed to increase cellular turnover while reducing the effects of free radical damage, allowing the skin to retain moisture and youthful suppleness (Thakur, 2011). Kale Extract is perfect for cosmetic and personal care products aimed at naturally revitalizing the appearance of dull, old, or damaged skin and hair.

Kava

Synonyms

Kava is a plant also known as Ava, Ava Root, Awa, Intoxicating Long Pepper, Kao, Kavain, Kavapipar, Kawa, Kawapfeffer, Kew, Lawena, Malohu, Maluk, Maori Kava, Meruk, Milik, Piper methysticum, Poivre des Cannibales, Poivre des Papous, Rauschpfeffer, Rhizome Di Kava-Kava, Sakau, Tonga, Waka, Wurzelstock, Yagona, Yangona, Yaqona, Yaquon, Yongona.

Kava is a beverage or extract that is made from *Piper methysticum*, a plant native to the western Pacific islands. The name "kava" comes from the Polynesian word "awa," which means bitter. In the South Pacific, kava is a popular social drink. It is consumed as a beverage in ceremonies to promote relaxation.

Uses

Kava has been utilized in alternative medicine as a potentially beneficial anxiety treatment.

Other unproven uses include cancer prevention, sleeplessness, depression, attention deficit disorder, reducing sedative withdrawal effects (from medications like Valium, Xanax, or Tranzene), and other disorders. A type of persistent anxiety marked by exaggerated worry and tension (generalized anxiety disorder or GAD). Several early studies show that taking kava doesn't improve symptoms in people with GAD (Stanisavljević et al., 2007)

It is unknown whether kava can be used to treat any medical problem. The FDA has not authorized this product for medical use. Kava should not be used in place of any medication that your doctor has prescribed for you.

Other unproven uses include cancer prevention, sleeplessness, depression, attention deficit disorder, reducing sedative withdrawal effects (from medications like Valium, Xanax, or Tranzene), and other disorders.

It is unknown whether kava can be used to treat any medical problem. The FDA has not authorized this product for medical use. Kava should not be used in place of any medication that your doctor has prescribed for you.

Kava affects the brain and other elements of the nervous system. Kavalactones, which are found in kava, are thought to be responsible for their effects. The active compounds of kava, known as kavalactones, are sticky, insoluble molecules. As it is ingested, it travels through the bloodstream, causing the plant to work as a muscle relaxant. The majority of research indicates that ingesting kava extracts with 70% kavalactones can reduce anxiety and may work as well as some prescription anti-anxiety drugs. Along with anxiety alleviation, some people take kava because they believe it improves their mood and makes it easier to fall asleep. Because of its calming qualities, it is occasionally used to treat attention deficit hyperactivity disorder (ADHD) and even drug withdrawal.

Side Effects

If you have any of the following symptoms of an allergic reaction, get emergency medical attention: hives; difficulty breathing; swelling of your face, lips, tongue, or neck (Nadda et al., 2020).

If you have any of the following symptoms, stop using kava immediately and contact your healthcare provider:

Nausea, upper stomach pain, itching, tiredness, loss of appetite, dark urine, clay-colored feces, and jaundice are all symptoms of liver disorders (yellowing of the skin or eyes).

Long-term kava consumption may result in major health concerns such as: dry, scaly, flaky skin; yellowed skin, hair, fingernails, or toes; red eyes, swollen cheeks; impaired ability to absorb protein; weight loss; lung difficulties; blood in your urine; or blood cell diseases that make it easier to bleed and fall sick.

Dosage and Extracts

Traditionally, people chew the roots of kava or grind them into a pulp and mix them with water. Kava is now accessible in a variety of non-traditional forms, such as pre-made liquid, tablets, and tea. It is also widely available in the form of tinctures (liquid extracts). Kava's flavor is frequently characterized as earthy and comparable to the soil. Because of the action of some kavalactones, some users may experience numbness in their lips and tongue for a few minutes after drinking kava. It is recommended that people consume it on an empty stomach to get the most out of it (Seth, 2003).

Dosages of Kava:

Suggested Dosing for:

Anxiety disorders

70% standardized extract: 100 mg orally three times daily

Kava-lactones: 60-120 mg/day orally

Root tea: 1 cup orally one to three times daily; 2-4 g root/150 ml water

Benzodiazepine withdrawal, prevention

70% standardized extract

Initial: 50 mg orally once/day, THEN

Titrate up over 1 week while tapering benzodiazepine over 2 weeks

No more than 300 mg/day titrated over 1 week

Insomnia

Kava-lactones: 180-210 mg orally at bedtime

Kava Kelp

Synonyms

Ecklonia cava is an edible brown alga that grows in warm coastal waters of the Pacific Ocean.

Though it has long been used as an herbal remedy for breastfeeding mothers and treatment for various health conditions, it has only recently become more popular outside of Asia.

Ecklonia cava grows atop rocks in large underwater forests along shallow coastlines in the Pacific.

In certain regions of Asia where seaweed and algae are a part of the diet, this brown alga may be eaten whole in soups and salads. In the West, Ecklonia cava is more common as a supplement in pill or powder form (Ningthoujam et al., 2012).

Much like other algae and seaweeds, Ecklonia cava has long been used as a traditional folk remedy in China, Japan, and Korea. In Japan, it's more commonly known as "kajima."

Ecklonia cava is rich in vitamins, minerals, protein, fiber, and other beneficial plant chemicals.

Uses

While Ecklonia cava is high in nutrients, not all of its traditional uses are supported by scientific evidence.

Furthermore, scientific research are still in its early stages. Because much of the available research was done in test tubes and on animals, the results may not apply to humans.

Here are some of the potential benefits of this alga that are best supported by research.

The plant is brimming with potent plant chemicals -

Ecklonia cava includes a variety of plant chemicals, including phlorotannins, which are thought to have antioxidant, anticancer, antidiabetic, and antiallergic activities.

According to one test-tube investigation, three of its phlorotannins, including one known as PPB (pyrogallol-phloroglucinol-6,6′-bieckol), exhibit exceptionally high antioxidant qualities.

PPB prevents DNA damage and has stronger antioxidant activity than vitamin C, according to laboratory studies.

Circulation may be improved.

According to several studies, Ecklonia cava improves circulation.

PPB supplements dramatically improved blood circulation, blood pressure, and some blood fat levels including cholesterol in a 4-week research in mice with obesity and high blood pressure (8Trusted Source).

Furthermore, a test-tube investigation discovered that PPB from Ecklonia cava may aid in the treatment of artery and vein dysfunction resulting from long-term poor circulation

- May reduce inflammation
- May fight obesity and other chronic diseases
- It May have anticancer effects
- May reduce your risk of heart disease

Figure 44: Kava Kelp.

Source: https://www.healthline.com/nutrition/ecklonia-cava

Dosage and Extracts

Ecklonia cava supplements are usually thought to be safe.

A 12-week human trial reveals that daily doses of up to 144 mg are safe and have no negative effects. These conclusions are supported by animal experiments.

Despite this, pregnant women should avoid Ecklonia cava due to a lack of safety studies. Although this alga has traditionally been used as a postpartum supplement, breastfeeding women should seek the advice of their healthcare professional before using it.

Ecklonia cava products are usually offered as pills, but they can also be powdered. Alginol and Seanol are two more names for this supplement.

Although there are no official dose recommendations, you should not exceed the dosage given on the box. The majority of marketed supplements contain 50–400 mg of Ecklonia cava extract per serving (Saslis-Lagoudakis et al., 2014). It's also a good idea to consult with your doctor before incorporating this product into your routine.

Side Effects

The utility of kava is still debatable. While some believe that kava can be used safely to relieve anxiety in the short term, others argue that the hazards outweigh the benefits.

Even when used for a short period of time, indigestion, mouth numbness, rash, headache, drowsiness, and vision abnormalities are common side effects.

Long-term use can have disastrous repercussions in some circumstances.

Kombucha Tea

Synonyms

Algue de Thé, Champagne of Life, Champignon de la Charité, Champignon des Héros, Champignon de Longue Vie, Champignon Miracle, Combucha Tea, Dr. Sklenar's Kombucha Mushroom Infusion, Kombucha Thé, Laminaire de Thé, Mushroom Infusion, Fungus Japonicus, Kargasok Tea, Kombucha Tea, Kombucha Mushroom Tea, Kwassan, Manchurian Fungus, Manchurian Mushroom Tea, Petite Mère Japonaise, Spumonto, T'Chai from the Sea, Té de Kombucha, Thé de Combucha, Thé de Kombucha, Tschambucco (Nazim et al., 2020).

Kombucha is created by combining tea (black, green, red, or white) with sugar and other ingredients and fermenting it at room temperature for 7-14 days with a specific yeast and bacteria mixture. It is sometimes wrongly referred to be a mushroom.

People drink kombucha to treat cancer, high blood pressure, and a variety of other ailments, but there is no clear scientific evidence to back up these claims.

Kombucha is made up of alcohol, vinegar, B vitamins, caffeine, sugar, and other ingredients. Kombucha may have antioxidant properties.

Uses

- Memory loss.
- Premenstrual syndrome (PMS).
- Joint pain (rheumatism).
- Aging.
- Loss of appetite.

- AIDS.
- Cancer.
- High blood pressure.
- Increasing white cell (T-cell) counts.
- Strengthening the immune system and metabolism.
- Constipation.
- Arthritis.
- Hair growth.
- Pain, when applied to the skin.
- Other conditions.

Side Effects

When drunk by mouth, kombucha tea can be dangerous for most adults. It can cause stomach issues, yeast infections, allergic reactions, yellow skin (jaundice), nausea, vomiting, headaches, and death.

Kombucha tea, particularly batches brewed at home in an unsanitary setting, can become infected with fungus (Aspergillus) and germs (including anthrax). Twenty people in Iran became infected with anthrax after drinking kombucha tea (Tuttolomondo et al., 2014). This tea is harmful to persons with compromised immune systems, such as HIV/AIDS patients, who are more susceptible to contracting infections, as well as when prepared in a lead-glazed ceramic kettle.

Kombucha tea is unsafe to consume during pregnancy and breastfeeding. To be on the safe side, avoid using it.

Kombucha tea contains alcohol, which can lead to alcoholism. If you have a drinking problem, stay away from it.

Diabetes: Kombucha tea may affect blood sugar levels in diabetics. If you have diabetes and drink kombucha tea, keep an eye out for indicators of low blood sugar (hypoglycemia) and closely check your blood sugar levels.

Diarrhea: Kombucha tea contains caffeine, which can cause diarrhea. The caffeine in kombucha tea, especially when consumed in excessive quantities, might aggravate diarrhea.

Irritable bowel syndrome (IBS): Caffeine is present in Kombucha tea. The caffeine in kombucha tea, especially when consumed in large quantities can worsen diarrhea and might worsen symptoms of IBS.

Surgery: Because kombucha tea appears to impact blood glucose levels, it is possible that it will interfere with blood glucose control during and after surgery. Stop drinking kombucha tea at least two weeks before your surgery (Peng et al., 2017).

Weak immune system: If you have a weaker immune system as a result of HIV/AIDS or other causes, avoid drinking kombucha tea. Kombucha tea can promote the growth of bacteria and fungus, which can lead to serious diseases.

Dosage

The optimal dose of kombucha tea is determined by various factors, including the user's age, health, and other conditions. There is currently insufficient scientific evidence to define an optimal range of doses for kombucha tea. Keep in mind that natural products are not always safe, and dosages can be critical. Before using, be sure to read the product label and consult your pharmacist, physician, or another healthcare expert.

Extracts

If you want to understand more about kombucha tea, don't confuse it with kombucha extract. The extract is fundamentally a concentrated version of the beneficial kombucha culture, also known as SCOBY, or symbiotic bacteria and yeast culture.

Choosing the Correct One

You may come across a variety of extracts, but it's critical to separate the good from the poor. High-quality kombucha extracts are frequently marketed as:

- One hundred percent pure
- Fresh kombucha mushroom cultures were used to create this product.
- Probiotics-rich
- There is no additional alcohol.
- Additive-free

However, one of the most essential considerations is the alcohol concentration of kombucha extract. Many vendors, mostly commercial corporations that have entered the "natural products" industry in reaction to the industry's increasing popularity, blend their extracts with alcohol (Rawat et al., 2020). This is problematic for a variety of reasons, the most important of which is that excessive alcohol is thought to destroy the beneficial probiotic bacterium and yeasts that make kombucha so appealing in the first place.

Krill Oil

Synonyms

Scientific Name(s): *Euphausia antarctica., Euphausia australis., Euphausia glacialis., Euphausia murrayi., Euphausia superba.*

Common Name(s): Arctic krill oil, Krill, Krill oil

Krill oil is the oil extracted from a tiny, shrimp-like animal. Krill is the primary food of baleen whales, manta rays, and whale sharks. The word "krill" means "whale food" in Norwegian. People extract krill oil, put it in capsules, and use it as medicine.

Uses

Krill oil includes fatty acids that are comparable to those found in fish oil. These fats are thought to reduce edema, cholesterol levels, and make blood platelets less sticky. Blood platelets that are less sticky are less likely to form clots.

- Dry eye. According to research, ingesting krill oil orally for around 3 months reduces symptoms of dry eye, such as redness and an unstable tear film.
- Depression. Early research indicates that consuming krill oil does not help depressive symptoms in teens.
- Diabetes. Early study indicates that consuming krill oil does not assist patients with diabetes to improve their blood sugar levels.
- Cramps during menstruation (dysmenorrhea). Early study suggests that consuming krill oil may alleviate menstrual cramp pain.
- High levels of cholesterol or other fats (lipids) in the blood (hyperlipidemia). According to some studies, ingesting krill oil lowers total cholesterol and LDL ("bad" or low-density lipoprotein) cholesterol while increasing HDL ("good" or high-density lipoprotein) cholesterol in persons with high cholesterol. Another form of blood fat, triglycerides, is also lowered. However, not all research agrees (Reinaldo et al., 2020).

- Hypertriglyceridemia is characterized by elevated levels of lipid triglycerides in the blood. In those with high triglyceride levels, taking krill oil twice daily for 12 weeks appeared to decrease triglycerides.
- Osteoarthritis. Early research indicates that ingesting 300 mg of krill oil daily decreases pain and stiffness in persons with osteoarthritis.
- Premenstrual syndrome (PMS) (PMS). Early study suggests that eating 2 grams of krill oil per day may help minimize PMS symptoms.
- Rheumatoid arthritis (RA). Early research indicates that ingesting 300 mg of krill oil per day decreases pain and stiffness in persons with RA.
- Skin aging.
- Alzheimer's disease
- Cancer.
- Coronary artery disease
- HBP
- Obesity.
- Parkinson's disease.
- Stroke.
- Other circumstances.

Generally, Krill oil is promoted as an alternative to fish oil supplements. However, only limited clinical studies with small sample sizes have been conducted.

Dosage

Most clinical studies have used omega-3 doses of 1 to 2 g/day as krill oil. Krill oil 500 mg capsules commonly contain eicosapentaenoic acid (EPA) 60 mg, docosahexaenoic acid (DHA) 30 mg, and astaxanthin 61 mcg.

For dry eye: Krill oil (Nutra-Life OceanClean red krill oil) providing 945 mg of eicosapentaenoic acid and 510 mg of docosahexaenoic acid daily for 90 days has been used.

Side Effects

When taken orally, krill oil is safe for most adults when used correctly for a short period (up to three months). The most common krill oil adverse effects are stomach upset, loss of appetite, heartburn, fishy burps, bloating, diarrhea, and nausea (Saslis-Lagoudakis et al., 2011).

Extracts

The whole krill is composed primarily of water (about 80%), lipids (up to 4%), protein (about 15%), chitin, and carbs. Phospholipids (40%), mainly phosphatidylcholine, long-chain polyunsaturated fatty acids including EPA and DHA, the antioxidant astaxanthin (also present in shrimp and crayfish), vitamins A and E, and the flavonoid 6,8-di-C-glucosyl luteolin are all contained in krill oil.

To yet, only a few clinical investigations with a small number of subjects have been conducted. The contrast between krill oil and fish oils as a source of n-3 fatty acid supplementation has not been thoroughly investigated, particularly when dose and EPA and DHA concentrations are equal.

L

Lavender

Synonyms

Alhucema, Common Lavender, English Lavender, French Lavender, Garden Lavender, Huile Essentielle de Lavande, Lavanda, Lavande, Lavande à Feuilles Étroites, *Lavande Anglaise, Lavande Commune, Lavande des Alpes, Lavande du Jardin, Lavande Espagnole, Lavande Fine, Lavande Française, Lavande Officinale, Lavande Vraie, Lavandula, Lavandula angustifolia, Lavandula dentate, Lavandula latifolia, Lavandula officinalis, Lavandula pubescens, Lavandula spica, Lavandula stoechas, Lavandula vera*, Lavender Essential Oil, Ostokhoddous, Spanish Lavender, Spike Lavender, True Lavender.

Lavender is a plant. Lavender flower and oil are used to produce medication.

Lavender is frequently used to treat anxiety, tension, and sleeplessness. It is also claimed to treat depression, dementia, pain after surgery, and a variety of other diseases, but there is little scientific evidence to back up many of these claims (Schultz et al., 2020).

Lavender is used as a flavoring agent in foods and beverages.

Lavender is utilized in the production of pharmaceutical items as well as as a fragrance ingredient in soaps, cosmetics, fragrances, potpourri, and decorations.

Lavender (*Lavandula angustifolia*) is frequently polluted with similar species such as *Lavandula hybrida* which is a cross between *Lavandula angustifolia* and *Lavandula latifolia*, from which lavandin oil is obtained.

Figure 45: Lavender can be distilled into an essential oil, and has a range of medical applications.

Source: https://www.medicalnewstoday.com/articles/265922#uses

Uses

Lavender oil is thought to contain antibacterial and anti-inflammatory properties that can aid in the healing of small burns and bug bites.

According to research, it may be beneficial in the treatment of anxiety, sleeplessness, depression, and restlessness.

According to some research, drinking lavender tea can aid with digestive difficulties such as vomiting, nausea, intestinal gas, upset stomach, and abdominal edema.

Lavender is used to treat pain from headaches, sprains, toothaches, and sores, in addition to stomach issues. It can also be used to keep hair from falling out.

- Alopecia areata is a condition that causes hair loss. There is some evidence that using lavender oil in combination with thyme, rosemary, and cedarwood oils may enhance hair growth by up to 44% after 7 months of treatment (Salim et al., 2019).
- Anxiety. According to some research, ingesting lavender oil orally for 6-10 weeks reduces anxiety and sleep and prevents anxiety recurrence in persons with mild-to-severe anxiety. Lavender, on the other hand, does not appear to be more helpful than the anti-anxiety medicine lorazepam (Ativan). So far, preliminary research contradicts the usefulness of utilizing lavender oil as aromatherapy to relieve anxiety.
- Canker sores. According to some studies, using two drops of lavender oil to the affected area three times per day will promote canker sore healing and minimize swelling and pain.
- Preventing falls. There is some evidence that securing a lavender oil-infused pad to the neckline of garments lowers falls among nursing home residents.
- After-Cesarean-Section Pain (C-section). According to certain studies, breathing lavender essence while taking pain relievers intravenously (through IV) can help lessen pain in women following a C-section.
- Cancer-related discomfort. According to several studies, utilizing lavender oil for aromatherapy massage does not relieve pain in cancer patients.
- Dementia. Lavender oil applied to clothes collars or used for aromatherapy massage does not appear to help mental function in patients with Alzheimer's disease or dementia.
- Complications following childbirth. According to the majority of studies, adding lavender oil to baths did not reduce pain in the area between the vagina and anus after childbirth. However, some evidence suggests that taking lavender oil baths quickly (within 12 hours) after labor may alleviate this pain (Sepúlveda-Cuellar et al., 2021).

Side Effects

When taken orally, lavender is considered safe for most adults in food levels. When taken by mouth at therapeutic doses, it is safe. Lavender may produce constipation, headaches, and an increase in appetite when taken orally.

When applied to the skin: When applied to the skin in medical proportions, lavender is harmless. It can occasionally cause inflammation, but this is uncommon.

When used as aromatherapy, lavender is safe to inhale.

Dosage

Orally

- Anxiety: For 6-10 weeks, 80-160 mg of a specific lavender oil component (Silexan, Dr Willmar Schwabe GmbH & Co. KG) was taken daily. For 8 weeks, 500 mg of dried, powdered lavender flowers were administered twice daily.
- Depression: For 6 weeks, 80 mg of a specific lavender oil component (Silexan, Dr Willmar Schwabe GmbH & Co. KG) was taken daily. For four weeks, 60 drops of diluted lavender tincture were given every day.

Inhaled

- Anxiety: For four weeks, 8 drops of an oil combination containing 2% lavender and rose essential oils were applied to a cotton pad and inhaled for 15 minutes twice weekly. Diluting 2 drops of an essential oil blend containing 75% lavender oil and 25% rose oil in 5 mL of jojoba oil yielded the lavender and rose oil formulation (Silva et al., 2010). For one month, 3 drops of lavender essential oil were applied to a cotton pad and inhaled for 5 minutes daily. 2 drops of lavender oil were put to a cotton pad and left by the pillow for one week. Three drops of lavender oil were put to a cotton pad and inhaled for 30 minutes. For 15 minutes, lavender oil was diffused around the waiting room.
- Depression: 8 drops of an oil blend containing 2% lavender and rose essential oils were applied to a cotton pad and inhaled for 15 minutes twice weekly for depression relief. Diluting 2 drops of an essential oil blend containing 75% lavender oil and 25% rose oil in 5 mL of jojoba oil yielded the lavender and rose oil formulation. For 8 weeks, 5 mL of aromatherapy oil including

lavender, sweet orange, and bergamot was massaged into the upper body twice weekly.

- Menstrual cramps: It has been used to apply 3 drops of lavender oil to the palms and inhale every 6 hours, or to place 3 drops of oil onto a piece of cotton and inhale for 30 minutes every day for the first 3 days of menstruation. During a monthly period, 2 mL of lavender oil applied to the belly for 15 minutes as part of an aromatherapy massage has also been employed.

Extracts

- Make the lavender. If you have fresh lavender on hand, cut the stems off the plant and hang them upside down or flat to dry. When dry, grind them in a coffee grinder to improve their texture before using.
- Pour enough alcohol over the lavender to completely cover it.
- If using dry herbs, use one weight part lavender petals and five weight parts alcohol - a 1:5 ratio for fresh herbs. 1:2
- To avoid corrosion, cover the jar with a plastic lid or place parchment paper between the metal lid and the jar. Allow it to infuse in a cool area for 4-6 weeks, shaking occasionally.
- Use the concentrated liquid after straining the herbal material through cheesecloth or a fine sieve free of any herbs.
- Use dark dropper bottles to store it. Don't forget to label it with the date you made it.
- Store in the dark, cold place for over 3-4 years.

Lecithin

Synonyms

Egg Lecithin, Lécithine, Lécithine d'œuf, Lécithine de Graine de Soya, Lécithine de Soya, Lecitina, Ovolecithin, Ovolécithine, Phospholipide de Soja, Phospholipide de Soya, Phospholipides de Soya, Soy Lecithin, Soy Phospholipid, Soy Phospholipids, Soya Lecithin, Soybean Lecithin, Vegilecithin, Vitellin, Vitelline (Teixidor-Toneu et al., 2018).

Lecithin is a lipid that is required by the body's cells. It's found in a variety of foods, including soybeans and egg yolks. Lecithin is a drug that is taken orally and is also used in the production of pharmaceuticals.

Acetylcholine, a molecule that conducts nerve impulses, is formed from lecithin.

Uses

Lecithin is used to treat memory diseases like dementia and Alzheimer's. It is also used to treat gallbladder illness, liver disease, certain types of depression, excessive cholesterol, anxiety, and eczema, a type of skin condition.

Some people use lecithin as a moisturizer on their skin.

Some eye medications may have lecithin as a component. It is utilized to maintain the drug in contact with the cornea of the eye.

- Gallbladder disease
- Dementia caused by Alzheimer's disease or other conditions. Taking lecithin alone, or in combination with tacrine or ergoloids, does not appear to improve cognitive ability in patients with dementia. It also appears to have no effect on the course of Alzheimer's disease.
- Cholesterol is high. According to limited studies, lecithin lowers cholesterol in both healthy people and people on cholesterol-lowering treatment (statins). Other research suggests that lecithin does not affect low-density lipoprotein (LDL) or total cholesterol levels in patients with high cholesterol (Thakur, 2011).
- Manic-depressive illness. Early study indicates that ingesting lecithin helps symptoms of mania such as delusions, garbled speech, and hallucinations.
- Dermatitis, dry skin Lecithin is frequently included in skin treatments to aid in moisture retention. People may tell you that it helps, but there is no valid clinical evidence demonstrating that lecithin is useful in this application.
- Athletic ability. According to a limited study, ingesting lecithin orally does not appear to boost athletic performance in trained athletes.
- Disorders of movement (tardive dyskinesia). Early research indicates that ingesting lecithin orally alone or in combination with lithium does not appear to alleviate symptoms.

- Parkinson's disease. Early research shows that 32 grams of lecithin daily does not improve clinical symptoms in people with Parkinson's disease.
- Stress.
- Anxiety.
- Eczema.
- Sleep.
- Other conditions.

Side Effects

Lecithin is safe for most people. It can cause some side effects including diarrhea, nausea, abdominal pain, or fullness.

Not enough is known about the use of lecithin during pregnancy and breastfeeding. Stay on the safe side and avoid use.

Dosage

The optimal dose of lecithin is determined by various factors, including the user's age, health, and other conditions. There is currently insufficient scientific evidence to identify an optimal range of lecithin dosages. Keep in mind that natural products are not always safe, and dosages can be critical. Before using, be sure to read the product label and consult your pharmacist, physician, or another healthcare expert (Stanisavljević et al., 2007).

Extracts

Lecithin can be extracted chemically or physically using solvents such as hexane, ethanol, acetone, petroleum ether, or benzene. It is commonly found in egg yolk, marine sources, soybeans, milk, rapeseed, cottonseed, and sunflower oil. It has a low water solubility but is a great emulsifier. Based on hydration and temperature, its phospholipids can form liposomes, bilayer sheets, micelles, or lamellar formations in aqueous solution. This produces a surfactant that is typically categorized as amphipathic.

Lemons

Synonyms

Bioflavonoid Complex, Bioflavonoid Concentrate, Bioflavonoid Extract, Bioflavonoïde de Citron, Bioflavonoïdes, Bioflavonoïdes de Citron, Bioflavonoids, Citrin, Citron, Citronnade, Citrus, Citrus Bioflavones, Citrus Bioflavonoid, Citrus Bioflavonoid Complex, Citrus Bioflavonoid Extract, Citrus Bioflavonoids, Citrus Extract, Citrus Flavones, Citrus Flavonoids, Citrus limon, Citrus limonum, Citrus Medica Limonum, Citrus Peel Extract, Citrus Seed Extract, Complexe de Bioflavonoïdes, Complexe de Bioflavonoïdes de Citron, Concentré de Bioflavonoïdes, Eriodictiol, Eriodictyol, Eriodictyol Glycoside, Extrait de Bioflavonoïdes, Extrait de Bioflavonoïdes de Citron, Extrait de Pépins de Citron, Flavonoids, Flavonoïdes, Flavonoïdes de Citron, Jus de Citron, Lemon Bioflavonoid, Lemon Bioflavonoid Complex, Lemon Bioflavonoid Extract, Lemon Bioflavonoids, Lemon Juice, Lemon Oil, Lemon Peel, Lemon Rind, Lemonade, Limón, Limonade, Nimbaka, Nimbuka, Zeste de Citron (Miara et al., 2018).

Uses

Lemon is a type of plant. Medicine is made from fruit, juice, and peel.

Lemon is used to cure scurvy, which is caused by a lack of vitamin C. Lemon is also used to treat the common cold and flu, as well as H1N1 (swine) flu, ringing in the ears (tinnitus), Meniere's illness, and kidney stones. It is also used to help digestion, relieve pain and swelling (inflammation), enhance blood vessel function, and stimulate urination to help with fluid retention.

Lemon is used as a culinary and flavoring element in dishes.

- Meniere's illness According to some reports, eriodictyol glycoside, a molecule found in lemon, may improve hearing and reduce dizziness, nausea, and vomiting in certain persons with Meniere's illness.

- Stones in the kidney. Not having enough citrate in the urine seems to raise the chance of getting kidney stones. Drinking 2 liters of lemonade throughout the day has been shown to considerably increase citrate levels in the urine. This may aid in the prevention of kidney stones in these individuals (Seth, 2003).
- Scurvy treatment. Scurvy is a vitamin C deficiency illness. Lemon can supplement vitamin C deficiency.
- Colds and flu are common ailments.
- Swelling is reducing.
- Urine production is increasing.
- Other circumstances.

Side Effects

Lemon is safe in dietary quantities and may be safe in greater therapeutic doses. Higher doses of lemon have not been studied for its potential adverse effects.

Applying lemon to the skin may increase the risk of sunburn, especially in persons with fair skin.

Precautions and Warnings: Pregnancy and breastfeeding: When consumed as part of a normal diet, lemon is safe for pregnant and breastfeeding women. However, it is unknown whether it is safe to utilize lemon at higher therapeutic doses during pregnancy or breastfeeding. Stick to the food portions.

Dosage

The optimal dose of lemon for treatment is determined by various factors, including the user's age, health, and a variety of other situations. At this time, there is insufficient scientific evidence to define a suitable range of dosages for lemon. Remember that natural products are not always safe, and dosages can be critical. Before using, follow the recommendations on the product label and consult your pharmacist, physician, or other healthcare expert (Saslis-Lagoudakis et al., 2014).

Figure 46: Lemon.

Source:

Extracts

Lemon extract is essentially that lemony taste you get from fresh lemons, but in a stronger, more concentrated liquid form. It's usually made from fresh lemons with an oil or alcohol base and the extraction process involves soaking lemon peels in a neutral spirit, like vodka.

Since the lemon extract is much more concentrated in flavor than lemon juice, use only half as much lemon extract and add back in the rest of the liquid in the form of water. For instance, if a recipe calls for 2 tablespoons of lemon juice, substitute 1 tablespoon of lemon extract and 1 tablespoon of water.

Lemon balm/Melissa

Synonyms

Balm mint; Blue balm; Garden balm; Honey plant; Melissa officinalis; Sweet balm

Lemon balm is a plant that originated in Europe and is now spread all over the world. It is planted not just in herb gardens or to attract bees, but also in crops for medicine, cosmetics, and the production of furniture polish (Nadda et al., 2020). If not managed, the plant can grow up to 2 feet tall, and occasionally even higher. Clusters of tiny, light yellow flowers grow where the leaves meet the stalk in the spring and summer. The leaves are severely wrinkled and vary in color from dark green to yellowish-green depending on the soil and environment. Your fingers will smell sour and sweet, like lemons, if you rub the leaves. The leaves resemble mint leaves in form and are from the same plant family.

Uses

Lemon balm (Melissa officinalis), a mint family member, is regarded as a relaxing herb. It has been used since the Middle Ages to relieve tension and anxiety, promote sleep, enhance appetite, and relieve pain and discomfort caused by indigestion (including gas and bloating, as well as colic). Lemon balm was brewed in a wine before the Middle Ages to lift spirits, heal wounds, and treat deadly insect bites and stings. To encourage relaxation, lemon balm is frequently paired with other calming, soothing herbs such as valerian, chamomile, and hops. It is also used to heal cold sores in creams (oral herpes).

Anxiety and insomnia

Several studies have shown that combining lemon balm with other relaxing herbs (such as valerian, hops, and chamomile) can help reduce anxiety and induce sleep. Except for topical application, few research has looked at lemon balm on its own. In one research of patients with modest sleep issues, for example, 81 percent of those who took a valerian and lemon balm herbal combination reported sleeping substantially better than those who received a placebo. According to this and other studies, it is unclear whether lemon balm or valerian (or a combination of the two) is to blame for the outcome (Ningthoujam et al., 2012).

Herpes

According to certain research, topical ointments containing lemon balm may aid in the healing of cold sores caused by the herpes simplex virus (HSV). In one research of 116 persons with HSV, those who applied lemon balm lotion to their lip sores saw significant reductions in redness and swelling after only two days. Other symptoms, such as discomfort and scabbing, remained unchanged. Lemon balm ointment was exceedingly effective, according to both the patients and their doctors.

Other applications

Some research suggests that lemon balm, when combined with other herbs, may aid in the treatment of dyspepsia. Others have discovered that lemon balm oil has potent antimicrobial properties. In one investigation, lemon balm was found to have acceptable antibacterial activity against Listeria monocytogenes and Staphylococcus auerus. A few studies have also discovered that lemon balm may assist improve cognitive performance and reducing agitation in persons with Alzheimer's disease.

Dosage

Consult a qualified physician for the particular dose that will best suit your needs if you are having trouble sleeping or reducing indigestion, gas, or bloating. The following are some possible doses:

Capsules: Take 300 to 500 mg dried lemon balm three times a day, or as needed.

Tea: 1.5 to 4.5 g (1/4 to 1 tsp) dried lemon balm herb in boiling water Steep and consume up to four times each day.

Every day, take 60 drops of lemon balm tincture.

Topical: Apply a topical lotion to the afflicted region three times per day, or as indicated.

Steep 2 to 4 tsp crushed leaves in 1 cup boiling water for 10 to 15 minutes to treat cold sores or herpes sores. Cool. Throughout the day, apply tea to the sores with cotton balls.

Side Effects

When used in food proportions, lemon balm is safe to take by mouth. Lemon balm is safe to ingest by mouth in medical levels for a short period. It has been used in a study for up to 6 months without incident. The most common side effects include increased hunger, nausea, vomiting, abdominal pain, dizziness, and wheezing. There isn't enough information about the long-term safety of lemon balm (Peng et al., 2017).

Lemon balm is safe for adults to use in medical doses when applied to the skin. It has the potential to irritate the skin.

Extracts

Figure 47: Lemon balm extract.

Source: https://www.amazon.com/Liquid-Extract-Melissa-officinalis-Tincture/dp/B00KLGBE4O

- Highest Strength, Super concentrated Lemon Balm liquid extract!
- Look out for Lemon Balm tincture that contains ONLY natural ingredients!
- High-Quality Lemon Balm liquid extract - crude extract of first extraction! We meticulously produce our extracts according to precise standards where each herb is extracted according to the distinct characteristic of each plant!
- Super concentrated Lemon Balm extract: dry material/menstruum ratio 1:3

Lemon Myrtle

Synonyms

Backhousia citriodora (common names lemon myrtle, lemon-scented myrtle, lemon scented ironwood) is a flowering plant in the family Myrtaceae, genus Backhousia. It is endemic to subtropical rainforests of central and south-eastern Queensland, Australia, with a natural distribution from Mackay to Brisbane. Other common names are sweet verbena tree, sweet verbena myrtle, (lemon scented verbena is another species), and lemon scented backhousia.

Uses

- Infection spread through sexual contact can result in genital warts or cancer (human papillomavirus or HPV). Placing myrtle leaf extract suppositories into the vagina on most days of the month aids in the treatment of HPV infections in women. It also appears to alleviate itching and burning (Nazim et al., 2020).
- Canker sores Early study indicates that using a paste containing myrtle leaf extract aids in the healing of canker sores.
- Heartburn that persists. Early research indicates that ingesting myrtle berry extract works just as well as the prescription omeprazole for relieving heartburn symptoms.
- Menstrual cycles with unusually heavy bleeding (menorrhagia). According to preliminary study, eating myrtle fruit syrup during menstruation helps to lower the amount of blood lost in women who have heavy periods.
- Bladder problems.
- Bronchitis.
- Diarrhea.
- Tuberculosis.
- The whooping cough.
- Worms.
- Other circumstances.

Dosage

For a sexually transmitted infection that can lead to genital warts or cancer (human papillomavirus or HPV): Suppositories containing myrtle leaf extract 10% and leaf essential oil 0.5% have been placed into the vagina on the non-menstruating days of each cycle for up to three menstrual cycles.

Side Effects

When applied on the skin or in the vagina, diluted myrtle leaf extract is harmless. It is possible to experience irritation and dryness. When consumed orally, myrtle oil is dangerous. It contains a chemical that can lead to low blood pressure, blood circulation issues, and other issues.

There isn't enough information to say whether eating a myrtle leaf, branch, or berry is safe.

Extracts

Lemon Myrtle essential oil is used as an antiseptic, disinfectant and expectorant and is said to be more powerful than Tea Tree. In aromatherapy, it is valued for its fresh scent with uplifting benefits on your emotions and the ability to soothe, your body and mind.

Figure 48: LEMON MYRTLE EXTRACT.

Source: https://aromaticingredients.com.au/products/lemon-myrtle-extract

Recognized by the aromatherapy industry for its uplifting freshness, extracts are ordinarily water-soluble actives are gaining recognition and featuring in soaps, lotions, lip balms and body butter, shampoos and shower gels; as well as a new dimension to existing Lemon Myrtle products.

Licorice Root

Synonyms

Acide Glycyrrhizique, Acide Glycyrrhizinique, Alcacuz, Alcazuz, Bois Doux, Bois Sucré, Can Cao, Chinese Licorice, Deglycyrrhized Licorice, East European Licorice, Gan Cao, Gan Zao, Glabra, Glycyrrhiza, Glycyrrhiza echinate, Glycyrrhiza glabra, Glycyrrhiza glabra typica, Glycyrrhiza glabra violacea, Glycyrrhiza glabra glandulifera, Glycyrrhiza glandulifera, Glycyrrhiza Radix, Glycyrrhiza uralensis, Glycyrrhizae, Glycyrrhizic Acid, Glycyrrhizinic Acid, Isoflavone, Jethi-Madh, Kanzo, Lakritze, Licorice Root, Liquiritiae Radix, Liquirizia, Mulathi, Mulethi, Orozuz, Phytoestrogen, Phyto-œstrogène, Racine de Réglisse, Racine Douce, Radix Glycyrrhizae, Régalissse, Regaliz, Reglisse, Réglisse, Réglisse Déglycyrrhisée, Réglisse Espagnole, Réglisse Russe, Regliz, Russian Licorice, Spanish Licorice, Subholz, Sussholz, Sweet Root, Turkish Licorice, Ural Licorice, Yashtimadhu, Yashti-Madhu, Yashti-Madhuka, Zhi Gan Cao (Rawat et al., 2020).

Licorice root, which is considered one of the world's oldest herbal remedies, comes from the root of the licorice plant (Glycyrrhiza glabra)

Native to Western Asia and Southern Europe, licorice has long been used to treat various ailments and flavor candies, drinks, and medicines

Uses

Eczema (atopic dermatitis). There is some evidence that applying licorice to the skin might alleviate eczema symptoms. Applying a licorice gel three times per day for two weeks appears to minimize redness, swelling, and itching.

Canker sores The majority of the study indicates that placing a licorice patch to the interior of the mouth reduces the size of canker sores but does not speed up healing time. Rinsing the mouth with licorice solution appears to help reduce pain and accelerate the healing of canker sores.

People's reactions to the removal of their breathing tubes. Sucking on a licorice lozenge or gargling with licorice fluid before inserting a breathing tube appears to help avoid coughing and sore throat from occurring when the tube is removed.

- Dental plaque.
- Dry mouth
- Menstrual cramps
- High cholesterol.
- Malaria.
- Muscle cramps.
- Pain in people with cancer.
- Weak and brittle bones (osteoporosis).
- Osteoarthritis.
- Prostate cancer.
- Scaly, itchy skin (psoriasis).
- Tuberculosis.

Figure 49: Licourice root.

Source: https://www.healthline.com/nutrition/licorice-root

Dosage

- For eczema (atopic dermatitis), gel solutions containing 1% or 2% licorice root extract were administered three times per day for two weeks.
- For canker sores, use: For 8 days, an oral patch containing 0.015-0.229 mg/kg licorice was applied 16 hours each day. A patch containing 1% licorice extract was applied. A mouth rinse containing 5% licorice extract was swished around the mouth for roughly 3 minutes four times per day until recovery was achieved. Warm water containing 200 mg of deglycyrrhizinated licorice powder was gargled four times per day for seven days (Reinaldo et al., 2020).
- Side effects in those who use a breathing tube: Before anesthesia, a lozenge containing 97 mg of licorice was sucked for 30 minutes. Gargling for at least one minute with 30 mL of a fluid containing 0.5 grams of licorice beginning 5 minutes before the installation of a breathing tube has been employed.

Side Effects

When ingested at the levels found in foods, licorice is generally safe for most people.

Glycyrrhizin-containing licorice products should not be consumed in high quantities for more than 4 weeks or in tiny doses over a lengthy period. Eating 5 grams or more of licorice every day for several weeks or longer can result in significant negative effects. Extremely high blood pressure, low potassium levels, weakness, paralysis, abnormal heart rhythms, and heart attack are all examples.

When applied to the skin: Licorice is safe when applied to the skin in the form of a gel containing 2% licorice root extract for up to 2 weeks or when used as a mouth rinse for up to one week. Some people may develop a rash as a result of it.

Lime

Synonyms

Adam's Apple, Bara Nimbu, Bijapura, Citron Vert, Citronnier Vert, *Citrus acida, Citrus aurantifolia, Citrus lima, Citrus limetta var. aromatica, Citrus medica var. acida*, Huile de Lime, Italian Limetta, Key Lime, Lima, Lime Oil, Limette, Limettier, *Limonia aurantifolia*, Turanj.

Lime is a citrus fruit that is sour, round, and brilliant green.

They're abundant in vitamin C, antioxidants, and other nutrients, making them nutritional powerhouses.

Limes come in a variety of varieties, including Key lime (*Citrus aurantifolia*), Persian lime (*Citrus latifolia)*, desert lime (*Citrus glauca*), and makrut lime (*Citrus hystrix*) (Saslis-Lagoudakis et al., 2011).

Each of these species has its own set of traits. The Key lime, for example, is smaller, more acidic, and fragrant than the more common Persian lime.

Persian limes are the most widely accessible in the United States.

Uses

- Low levels of healthy red blood cells (anemia) due to iron deficiency.
- Malaria. Early research shows that taking lime juice along with standard malaria medicine may help treat children with malaria better than taking malaria medicine alone.
- Sickle cell disease.
- Quitting smoking.
- A disease caused by vitamin C deficiency (scurvy).
- Dysentery.
- Nausea.
- Killing germs on the skin.
- Vaginal infections when used in the vagina.
- Other conditions.

Dosage

The optimum dose of lime is determined by various factors, including the user's age, health, and other situations. There is currently insufficient scientific evidence to identify a suitable range of lime dosages. Keep in mind that natural products are not always safe, and dosages can be critical. Before using, be sure to read the product label and consult your pharmacist, physician, or other healthcare experts.

Side Effects

When consumed orally, lime is safe for most adults in concentrations found in foods. There isn't enough credible information to tell whether lime is safe to use as a medicine or what the potential adverse effects are (Schultz et al., 2020).

When applied to the skin: It is not safe to apply lime directly to the skin. When lime is applied directly to the skin, certain persons are sensitive to it. Lime can make your skin extremely sensitive to sunlight. Outside, wear sunscreen and protective clothes.

When used in the vagina: Using lime juice in the vagina is dangerous. Lime juice can be damaging to vaginal and cervix cells. Itching, burning, dryness, discomfort, and other symptoms may occur.

Extracts

Lime extract has a zesty lime flavor that is used most often in baking and beverages. The lime flavor in this extract comes from distilled lemon oil pressed from the skin of limes. Just a teaspoon will add a little twist to fudge, salsa, dip, or drinks.

Liverwort

Synonyms

American Liverleaf, Anemone acutiloba, Anémone à Lobes Aigus, Anemone americana, Anemone hepatica, Anémone d'Amérique, Anémone Hépatique, Bryopsidée, Hepatica nobilis var. acuta, Hepatica nobilis var. obtusa, Hepática, Hepatici Noblis Herba, Hépatique d'Amérique, Hépatique à Lobes Aigus, Herb Trinity, Kidney Wort, Leberbluemchenkraut, Liverleaf, Liverweed, Liverwort-Leaf, Round-Leaved Hepatica, Round-Lobe Hepatica, Sharp-Lobe Hepatica, Trefoil (Tuttolomondo et al., 2014).

Uses

Despite significant safety issues, liverwort is used to treat gallstones and liver diseases such as jaundice, liver enlargement, hepatitis, and liver cirrhosis. It is also used to treat stomach and digestive tract discomfort, to stimulate appetite, to relieve fullness, to regulate stool function, and stimulate the pancreas.

Some individuals use liverwort to heal varicose veins, decrease cholesterol, increase blood circulation, and "purify" blood.

Women use liverwort to alleviate menopausal symptoms.

Other applications include nerve strengthening, metabolic stimulation, relaxation promotion, and use as a general tonic.

Dosage

The optimal dose of liverwort is determined by numerous factors, including the user's age, health, and other conditions. There is currently insufficient scientific evidence to define an acceptable range of dosage for liverwort. Keep in mind that natural products are not always safe, and dosages can be critical. Before using, be sure to read the product label and consult your pharmacist, physician, or other healthcare expert (Salim et al., 2019).

Side Effects

When consumed orally, fresh liverwort is harmless. It can cause diarrhea, stomach inflammation, renal and urinary tract irritation, and other side effects. There isn't enough credible information to establish whether dried liverwort is safe or what the potential side effects are.

Fresh liverwort is dangerous when applied to the skin. It can cause itchiness, itching, and pus-filled blisters. There isn't enough credible information to establish whether dried liverwort is safe or what the potential side effects are.

Extracts

The chemical composition of three different extracts of the liverwort S. nemorea was determined by solid-phase microextraction gas chromatography-mass spectrometry (SPME GC–MS). The dominant compounds in extracts were sesquiterpenes β-bazzanene (11%, 17.9% and 14.6%), isobazzanene (10.2%, 15.8% and 11.7%) and aromadendrene (8.8%, 12.9% and 10.6%) in the methanol, ethanol and ethyl acetate extracts, respectively. The antimicrobial potential of S. nemorea extracts against food spoiling yeasts and bacteria was determined in vitro by the microdilution method.

Lobelia

Synonyms

Asthma Weed, Bladderpod, Emetic Herb, Gagroot, Herbe à Asthme, Indian Tobacco, Lobelia inflata, Lobélie, Lobélie Brûlante, Lobélie Enflée, Lobélie Gonflée, Pukeweed, Tabac Indien, Vomit Wort, Wild Tobacco.

Uses

Lobelia is used to treat newborn infants with breathing disorders such as asthma, bronchitis, whooping cough, and shortness of breath (apnea). Lobelia is sometimes used as a sedative to help individuals relax. Others use it to make them sweat more.

Lobelia is applied to the skin to treat muscle discomfort, rheumatoid arthritis joint lumps (rheumatic nodules), bruises, sprains, insect bites, poison ivy, and ringworm.

Lobelia is utilized in cough treatments and anti-irritant medications in the pharmaceutical industry. Lobelia is a component in some stop-smoking products around the world (Sepúlveda-Cuellar et al., 2021). However, manufacturers have been prohibited from using lobelia in stop-smoking medications sold in the United States since 1993. That's when researchers found that lobelia doesn't make stop-smoking products any more effective.

Dosage

The optimal dose of lobelia is determined by various factors, including the user's age, health, and other conditions. There is currently insufficient scientific evidence to define an acceptable range of dosages for lobelia. Keep in mind that natural products are not always safe, and dosages can be critical. Before using, be sure to read the product label and consult your pharmacist, physician, or other healthcare expert.

Side Effects

When used orally, obelia is deemed dangerous for the majority of people. Nausea, vomiting, diarrhea, cough, disorientation, tremors, and other significant side effects are possible (Teixidor-Toneu et al., 2018).

Overdoes can result in a variety of dangerous toxic consequences, including sweating, convulsions, rapid heartbeat, extremely low blood pressure, collapse, coma, and death. It is reported that 0.6-1 gram of the leaf is toxic, and 4 grams might be lethal.

M

Marvel Grass

Marvel grass isa species of grass that is mostly used as a forage for livestock. This grass is normally considered to be tropical given the fact that it is native to the North African and Indian region. Individuals from the tropical and sub-tropical zones use this kind of grass for pasture. The Indian region is known to mostly use this grass. Many names are used in the description of this grass. Such include marvel grass, hindi grass, sheda grass, ringed dichanthium, vuda blue grass, bluestem, diaz bluestem, Santa Barbara grass [English]; karad [English/India]; pitilla, kleberg blue stem, jargu grass, Delhi grass, two-flowered golden-beard, climacuna, yerba de vias is the name the Spanish use to describe it; vleivingergras as described by the Afrikaans and 双花草 as described by the Chinese (Asnake et al., 2016).

Scientifically, this grass is normally described as *Dichanthium annulatum* (Forssk.) Stapf [Poaceae]. Some of the synonyms of marvel grass include Andropogon papillosus Hochst. ex A. Rich, *Dichanthium nodosum* Willemet, *Dichanthium papillosum* (Hochst. ex A. Rich.) Stapf or *Andropogon annulatus* Forssk. Marvel grass is known to be a tufted perennial crop that has a height of approximately 80 cm. its root system is very important and grows up to 1m deep. In the case of its clumps, it is considered to be erect and decumbent to a size of 25-100 cm long. Its leaf blades are 2-7mm wide , 3-30cm long and are considered to be linear. These leaves are also pubescent on the margins. The inflorescences present in this grass are known to be composed of 2-15 racemes. Each of the racemes are 3-7m long. The grass also contains seeds that are 2mm long and described as an oblong-obovate caryopsis.

Figure 50: Marvel grass.

Source: image by feedipedia

Given the fact that this grass is much appreciated by all classes of ruminants, it is an excellent and widely used fodder grass. In the case where there exist mixed pastures, marvel grass is widely appreciated to all the other grasses. This product can be utilized in several ways including in pastures, for hay and silage making especially if it is cut before it flowers and for a cut and carries system where the pastoralists cut the grass and take them to their animals. It is recognized that in India marvel grass has a high capacity to withstand heavy grazing.

This is a homegrown medicinal plant that is used in the treatment of many diseases including high blood pressure, high cholesterol and other cardiovascular diseases, diabetes, appetite disorder, rheumatoid arthritis and hot flush.in New Zealand, the medicinal value of marvel grass has been recognized for a very long time. Apart from having medicinal value, they have been used as cuisines. It is a USDA-verified organic medicinal plant. Apart from being able to heal the physical conditions mentioned, it is said to be a full spectrum medicine that heals the spirit, nourishes the soul, perfects the emotions, quiets the mind, and strengthens the will. For the native individuals living in the Amazon region, this plant is known for healing wounds, fractures and internal injuries (Douwes et al., 2008). In addition, it has properties that boost your energy naturally, lessens pain and calms down your nerves, kills bacteria and fungi, helps with spasms of the muscles and joints, protects the brain from the negative impact of bacteria, removes excess salt from your body, strengthens the heart muscle and blood vessels, improves your complexion and even tastes great in a salad or in foods.

There is no information regarding the use of marvel grass and thus, it is important to follow the prescription of a physicist and consume it in small quantities.

Mahua

Figure 51: Mahua plant (Madhuca Longifolia).

Source: image by Feedipedia

Mahua is a plant native to tropical and sub-tropical regions of South America. This plant is scientifically described as Madhuca longifolia which belongs to the family sapotaceae, is known for its sweet flowers. This hardy plant survives drought and cold temperatures, as well as heavy winds. When taken from its natural habitat it can grow outdoors in temperate climates. Mahua is a tough plant that grows easily, re-seeding itself year after year to create a lush garden. This is a plant that can be found in tropical zones. It is named after Mahua Naga, the Bengali saint of snakebite. The mahua plant is well known for its soft leaves, which grow in a zig-zag pattern along its stalk. the plants can grow to heights of between 1 meter and 2 meters (3 to 6 feet) depending on the variety. Their shape is generally "prickly" or dense, which many people consider attractive in a houseplant. This mahua (Bassia latifolia) was one of the first trees to be planted in Brooklyn's Prospect Park. It has a very twisted trunk and a canopy that spreads out at the top so wide, birds have been known to build nests there.

This plant is known to possess many ethnic values among different tribal individuals around the world. These tribal individuals are known to be able to create fermented and non-fermented products from the plant. The non-products include halwa, barfi and methi puri. Some of the fermented products that are made from this plant include the mahua daru or the mahuli.

Usage

mahua (pronounced mow-hoo-ah) is a plant that produces woody stem sticks used in furniture construction. The common mahua tree, whose scientific name is Clusia, is found in the moist lowlands of South America. It has been cultivated since ancient times for its fruit and as a source of traditional medicine. Part of the sunflower family, mahua provides high levels of antioxidants and essential fatty acids – especially omega-3 – that can be beneficial to your brain and heart. It's also packed with nutrients like vitamin C and calcium, and the seeds provide a great source of protein and polyunsaturated fats, making it an excellent food for preventing cardiovascular disease (Do et al., 2019). Furthermore, the mahua can be used to purify water from nearly any source. It removes bacteria, viruses, and other harmful contaminants from water, while preserving healthy minerals and electrolytes, leaving users with a tastier, healthier drink. The plant's seeds contained valuable oils used to flavor food, lubricate machinery, and even fuel lamps. After extracting the oil, farmers would toss the seeds aside. Later, ecologists discovered that the seeds have valuable properties that have helped restore ravaged forests.

Given the above usages, the most important thing s the fact that the mahua plant is normally used as medicine for a variety of diseases. Such include eye diseases, diarrhea, headache and skin diseases. The leaves are used to fight against cancer, tuberculosis, cough and diabetes. Mahua flowers are used in curing acne. Mahua mixed with other ingredients is also an effective tonic for diarrhea and diabetes, reduces phlegm and inflammation of lungs etc. Mahua oil massages are good for improving eyesight and getting relief from sleeping problems. In China, it is known to be medicine described as Mahua Shen Jun. It has a various medicinal effects, such as antibacterial, antifungal, improving blood circulation and regulating excessive menstruation. It also can be a unique food for daily life.

In other instances, Mahua is a natural antibiotic, and painkiller and can help cure diseases like cancer, and asthma among other illnesses such as kidney stone, stomach pain, gall bladder stone, blood pressure, diabetes and fever. mahua made of the roots of the mahua shrub is most commonly used in Ayurvedic medicine. It is most known for its antifungal, antibacterial, and anti-inflammatory properties. It is most commonly used to treat wounds, tumors, respiratory issues, ulcers, eczema, diarrhea, and dysentery. Some of the other advantages of the Mahua plant include Soothing Pain, Disperses Cold, Clears Heat, having good effect on the spleen, stomach and lung inflammations. Its powder can be applied on boils or other skin swelling to alleviate pain. Its leaves can be inserted into festering or swollen wounds to disperse Cold and clear Heat for quicker healing.

Mentha Arvensis

Figure 52: Mentha arvensis.

Source: image by Madrean Discovery Flora and Fauna Database

This is a species of flowering plant that belongs to the mint family. It can be described in many ways which include field mint, corn mint or wild mint. It belongs to the mint family Lamiaceae which has a circumboreal distribution. This plant is native to Europe, Asia, and North Africa. It is most commonly cultivated in temperate zones around the world for its menthol flavoring agent, which is extracted from the leaves. The active ingredients are menthol, (-)-menthone, (-)-pulegone, menthyl acetate, menthofuran, limonene, linalool (Cummings & Olsen, 2011).

The plant has a long history of use in medicinal and cosmetic products. It is extensively cultivated in Europe and North America for its characteristic flavor to peppermint oil. Spearmint is used for many conditions including flatulence, colic, bronchitis, sore throat, cough, fevers and to alleviate feeling sick. It can be ingested in the form of tea or applied to the skin in the form of lotion. With a long history of use for coughs and respiratory complaints, M. arvensis is a traditional medicinal herb that is valued for its ability to soothe and promote a peaceful night's sleep. The fresh, minty aroma and flavor of peppermint add an iconic feel to everything from juleps and pairings with scampi to salads and desserts. This versatile mint is easy to grow and thrives in almost any soil type. It's grown indoors in pots and makes a lovely houseplant.

The dosage of this product is normally dependent on many factors which include the ser's health, age and several other conditions. Currently, there exists insufficient information regarding the right dosage of the product. As such, individuals are required to consult with their physicians at the event they decide to use the product.

Even though this plant is known to be beneficial, there are various possible side effects. When consumed through the mouth, the mint oil from the plant may be safe to most individuals but in other instances, it may cause stomach upsets. Some have allergic skin reactions when they apply the same to their skin. For individuals who have asthma, it can get worse if they get to inhale the product.

Mimosa Pudica

Mimosa pudica is derived from a Latin word known as pudica which means, shrinking or bashful. It is also called the sleepy plant, the sensitive plant, the sensitive plant, the shame plant or touch-me-not. It is a perennial flowering plant or otherwise a creeping annual plant that belongs to the legume family Fabaceae. *Mimosa Pudica* is the perfect home decor! The plant has Bold green variegated leaves that cover the branches of this houseplant (Carvalho & Morales, 2010). The leaves fold together when touched, making it an unusual and amusing conversation piece. The Mimosa Pudica plant is a lush, fast-growing plant that is loved for its enchanting and mysterious way of closing its leaves in the presence of a threat or disturbance. Easy to grow, this plant is popular for its ability to survive in all kinds of conditions including hot weather.

Figure 53: the sensitive plant.

Source: image from the spruce

The *Mimosa Pudica* is also known as, "The Shy Plant" due to its reaction when touched. When agitated, the leaves fold inward and the plant appears dead--only to spring back to life when touched or otherwise stimulated after some time has passed. This is a gift that will be used for many years to come--and one that will amaze your friends. This charming, touch-sensitive plant is native to the tropics of South America. A few mimosa leaves will lie flat in response to being touched, but when the plant is moved from a sunny area to a dark place the leaves will change from green to a rich bluish color. To repeat this dramatic effect, move the pot back into a sunny location and wait for a few days or weeks for the leaves to return to their original color. Growing up to 10 feet tall, this creeping vine bears clusters of delicate bell-shaped flowers and does a wonderful job of climbing a stake or trellis. Ripens up to 2 weeks before other fruit trees. Produces 3-4 fruiting cycles each year, maximizing harvest. Makes an attractive full-growing plant. The most pungent crab apple. Perfect for small yards, patios, balconies, bonsai.

This is an excellent plant to grow if you are interested in organic medicine. Its seeds can be ground down and mixed with other plants to make teas, tinctures, poultices, salves, oils, creams or balms. Its tiny pods contain the edible heart-shaped seeds prized by Native Americans for its medicinal properties. Mimosa was used traditionally for healing wounds, snake bites, and mild forms of arthritis in the hardy in zones.

This product is known to be available from commercial manufacturers in which the most common dosages are in the forms of capsules and liquids. the suggestion provided by manufacturers include3 to 6 mL of 1:2 extract that is in liquid form and 1 capsule 3 times a day daily with meals. The other extractions available from the plant include the tinctures, powders and teas (Bonini et al., 2018).

The possible side effects of this plant can only be seen with the case of individuals that have allergic reactions to this product. Given the fact that there is no information regarding the safety of the product with the case of pregnant and lactating omen, it is recommended that they do not use the same.

Michelia Champaca

Michelia champaca is commonly described as magnolia champaca. It is a large evergreen tree that belongs to the family magnoliaceae. The various synonyms include Hemapushpa, Champeya , Hemavha among others. It is commonly known for its woodworking properties and for its fragrant flowers. The large tropical leaves of the *M. champaca* plant are iconic, and for centuries it has been cultivated in Asia to perfume bedrooms. It soothes restlessness and eliminates stress, making this an ideal houseplant for busy families with young children (Braun, 2005). The *M. champaca* is a spectacular evergreen tree from the Asian tropics with a large, umbrella-like canopy of dark green leaves, and an unforgettable fragrance that comes from tiny flowers that grow in spring. The fragrance used to be bottled up and sold as essence of magnolia. The flowers are small and yellow.(An interesting note: the wood underneath the bark was once used to make chess pieces.). It is native to subtropical and tropical areas of southern China, where it grows in the wild. However, it's also cultivated in many other parts of the world for its scent and beauty, and can grow indoors year-round in subtropical climates, including Louisiana.

This plant produces sweet-smelling flowers that bloom in the winter and are often used for their scent, which is believed to have positive health benefits. The flowers can be distilled to make essential oils. For the case of the Indonesians and Philippines. The plant has been highly prized for its flower buds and distilled oils, which it has been highly sought after for centuries. Apart from the medicinal properties, the plant is known to have spectacular wood used for making violins, guitars, and other musical instruments. In traditional medicine, the bark is used to treat snake bites and malaria. The flowers are also used to treat fever and inflammation. There are Aromatic *M. champaca* is essential for making the classic French perfume, Chanel №5. Its sweet vanilla orchid-like scent is long-lasting and blends well with fresh florals, musky woods, and grassy greens. In Hawaii and Florida, the plant has become neutralized. The *M. champaca* is also known as Ceylon cinnamon, and in the old days was used for culinary and medicinal purposes.

Figure 54: the flowering of magnolia champaca.

Source: image by Online shopping store in Guatemala

The plant can be made in powdered form or a decoction made from the bark. The dosage of the powder is 3-6 g and that of the decoction is 50-100ml.

Current studies show that there are no known side effects of this plant and lactating mothers can use the same. However, in the case of pregnant women, there is need to consult a physician before deciding to use the same. Research has also shown that this plant or the extracts can be used in conjunction with other medicines.

Mustard

There are different types of mustard which include the white/yellow mustard, the brown mustard and the black mustard. In most instances, the whole, ground, bruised or cracked mustard is normally mixed with lemon juice, water, vinegar or even other liquids, salt and even some flavoring spices to create some form of paste. The mustard plant is a common garden all-purpose plant that produces green leaves and yellow flowers. The mustard plant was most likely domesticated between 5,000 to 6,000 years ago. The most notable use of mustard seeds is in making mustard, a popular condiment for meats and vegetables (Bhattacharyya, 2015). The Odorless Mustard Plant is one of the non-toxic plants that can be used as an indoor plant, although it also does well outdoors as a potted plant. It is a beautiful plant with small yellow flowers and feathery leaves, and requires very little care. The leaves can be used as an herbal remedy for warts. The seeds are known to have a sweet smooth flavor. In other instances, it is considered to be a useful condiment in making hot dogs especially when the customer requests spicy food. As a matter of fact, this spicy yellow wonder will set off the flavor of any plain old dog.

Figure 55: Mustard plant.

Source: image from Wikipedia

In addition, Mustard is a medicinal plant native to southeastern Europe and western Asia. It is grown as an annual plant, primarily for its seeds, which are used as a spice or a preservative. It is used to produce compound with interesting pharmacological properties, also known as mustard oil. The medicinal properties of mustard were known in ancient times. Mustard oil is commonly used in ayurvedic medicine and in Arabic traditional medicine. The use of mustard was mentioned in the bible already over 3000 years ago. For over 4000 years, the mustard plant has been recorded to have therapeutic properties (Blancke, 2016). The herb has been in use for various purposes such as relieving aches and pains, to preventing heart disease, improving circulation, removing phlegm, stimulating the appetite & helping alleviate hangovers. it is believed that its main therapeutic effects are due to its anti-inflammatory, analgesic, and antimicrobial properties. The herb contains various active components such as curcuminoids (diferuloylmethane) that are responsible for pain relief and anti-inflammatory effects. Other biological activities that are performed by the plant include antiproliferative, anti-leukemic, antibacterial, antiviral, antispasmodic, and wound healing properties among others. In other instances, the plant is used to relieve an individual from symptoms such as colds, fever, headaches or coughs.

Most mustards are likely to be safe when they are consumed as food but there is limited information regarding the various side effects. This especially concerns the consumption of the same through the mouth or even its application to the skin. There are however some known side effects which include diarrhea, heart failure, breathing difficulties, drowsiness, coma and in some instances even death. With the case of black mustard, one shouldn't apply it on the skin for a long time as it can be able to cause skin blisters and skin damage. In the case in which one is pregnant or breastfeeding, the black mustard is not the product for you and the rest of the mustard better be used with the aid of a physician.

The dose of mustard that is considered to be appropriate normally depends on a number of factors including the user's health, age and the other conditions that he/she might be facing. It should always be kept in mind that most natural products may be quite harmful and thus dosage is a very important aspect when it comes to dealing with its consumption.

Murraya Koenigii

Figure 56: Murraya Koenigii.

Source: image from PI@ntNET

There are vibrant colors in bougainvillea are capable of showing the beautiful story of this particular plant. These plants are known to normally flower during the winter and are known to be priceless given their showy blooms that occur year after year. There are also flowering vines that grow as low as 3 feet and there are those that grow as far as 20 feet and are known to produce thousands of flowers. To slow down the pace of life in your home with a beautiful accent plant. Murraya koenigii is known for its fragrant flowers and glossy green leaves (Bussmann & Sharon, 2006). Paired with a simple round pot, this plant offers a contemporary display piece that's easy to care for and suitable for home or office. This plant also known as the Curry leaf tree is a plant of the family Rutaceae. It is a tropical and sub-tropical tree native to India, Bangladesh, Pakistan, Sri Lanka, Thailand, Vietnam, Malaysia, Indonesia, Jamaica, and Burma . It grows up to 9 meters (30 ft) tall. Its fruit is an edible drupe. The plant is known to be growing thickly thorny stems of 0.5 to 2 cm thick, with thorns of 25-60 mm long. The leaves are alternately arranged along the branches of petioles of 3-5 mm long, oval or oblong-lanceolate, thick dark green.

This plant has similar qualities as the Jequirity Bean plant. It is described as a short perennial, shrub that normally grows in most parts of Africa. Each part of the plant is useful for its medicinal properties. The dried seed pods are ground into a fine powder and used for traditional medicine in India, which has led to further uses. Commonly, it is known as koyy koni.

The herb pod of *Murraya koenigii* Rosc. was initially used for the treatment of breast cancer, bile duct cancer, lung cancer, cervix cancer and bladder cancer. Most individuals believe that this plant can be used to create many different medicinal formulations because of its rutinoside, tannin, saponin, flavonoid, flavonol, proanthocyanidine glycosides, polyphenols and vitamin C. apart from this, it is said that the plant is also a good ayurvedic drug useful in maintaining health. The leaves of the plant are known as Curry Leaves, are used as an essential ingredient in curries, imparting strong flavor to the dish, while also having antifungal, antiviral and antibacterial properties. Curry Leaves are commonly used in Southern Indian states like Andhra Pradesh, Tamil Nadu, Kerala and Karnataka. Most individuals use curry extracts in the flavoring of food. the plant has striking leaves and very large flowers that are in the same genus as the cardamom which shows that the plant has good aromatic features. It is said that the beauty of the plant has been coveted over the centuries (Agbogidi & Azagbaekwe, 2013).

The recommended dosage of the curry leaves is ½-1 teaspoon. Otherwise, individuals may as per their recommendations. Even though the curry leaves have numerous medicinal and culinary benefits, it is no doubt that there also exist some possible side effects. The leaves are normally tinged with some form of sourness and as such may not be able to taste the best. In excess consumption, it may *produce some possible side effects such as stomach upsets, acidity in people and even nausea.*

Myristica Fragrans

This is known to be an evergreen plant that is native to the individuals of Moluccas commonly described as the spice islands. it has culinary benefits for individuals in the region as it produces nutmeg and mace. This species of plant is known to be dioecious as it consists of the male system and the female which are its flowers or the carpelliate flowers. The flowers of the plant are normally bell-shaped somewhat waxy and fleshy and also pale yellow. They also contain white flowers that are followed by fruit that can be used for a variety of culinary dishes. In most cases, they are staples and can be found within the home garden and are known to be easy to be cared for. These plants are tender and are very hard to trace in the united states.

It is considered to be a rare species of plant and the seeds are used for different medicinal purposes. The nutmeg spice has a delicious flavor but it should be used in small amounts since it induces sleepiness and dizziness and as such individuals are refrained from driving and handling heavy machinery at the point they consume this product. The seeds of the plant were historically considered an important trade good. It is used as a spice to flavor food or as a healing remedy. The plant has a round fleshy fruit (mace) resembling an apricot. The fruit shell (aril) is used as a spice, while the seed kernel (nutmeg), is used as a spice and a medicine. Nutmeg oil; which can be extracted from nutmeg fruits; has long been used in perfumes and flavorings.

Figure 57: Myristica fragrans.

Source: image by IPlantz

Given the fact that the plant has various medicinal advantages, it does not lack some side effects. If consumed in amounts that are normally found in food, then possibly, there might not be any side effects. Otherwise, if taken in amounts that are higher than those that are normally present in foods, then it may be unsafe for the consumer. There are cases where individuals that have taken high doses of the plant are known to experience dry mouth, nausea, irregular heartbeat, agitation, dizziness and hallucinations. In most serious cases, it has been known to cause death. There is limited knowledge regarding the side effects of the same on the skin when physically applied (Athar & Nasir, 2005).

The dosing of the product is normally dependent on the user's age, health among other conditions. There is currently limited knowledge regarding the appropriate dosage of mace and nutmeg. As such it is only relevant that an individual follows the instructions provided by the physician.

N

Nicotiana Tabacum

Given the fact that this is an easy-to-grow plant, it has the capability of growing in most parts of a country. Nicotiana is native to the Americans and has many different colors and flowering combination. Very little care is needed and can be able to withstand a range of conditions. This plant is commonly described as tobacco and is an herbaceous plant growing annually. This plant is native to individuals living in the Caribbean and the first individuals to use it were the Taino people. The other names of the plant include Nico Tabacum Tobacco Plant, NicoFlameTM, Zippo, Zippo Lighter. This plant holds the key to the tobacco flower. The so-called flower of the gods in all its diversity demonstrates not only its extraordinary beauty but also its multiple uses. The plant has delicate clusters of flowers that have been part of the rituals of Native Americans for centuries currently been developed into a fresh tobacco plant for all to enjoy. In addition to the tropical tobacco plant, the variety pack contains seeds for four additional strains of tobacco. If organically grown, it can be used in one's favorite pipe or even cigarettes. This plant belongs to the family Solanaceae, best known for being a major source of tobacco. It is a close relative to the tomato and potato plants and is edible in its entirety. When in the wild the plant can grow up to a height of 180 centimeters but when it is cultivated the height can go up to 4 meters.

Figure 58: Nicotiana tabacum.

Source: image by www.kalliergeia.com

There are normally calmer and milder versions of the plant that can be easily grown and have little to no odor than regular tobacco. It is a shame that once different cultures began using the leaves, it introduced an addiction to cigars, pipes and cigarette smoking which now kills millions worldwide.in a year, the plant produces thousands of seeds containing up to 40 percent nicotine.to prevent the development of mold, the plant needs to be dried for two weeks or even more. With the formation of mold, the nicotine content would be reduced. It is known that tobacco can be able to help with heart attacks, strokes, asthma and stress. In other instances, seeds are smoked, embedded in alcoholic solutions made with sugar, or rubbed on skin diseases. This plant is also known to relax frazzled nerves, dry up mucus in people who suffer from chronic coughs and has a long-lasting height that individuals can be able to relax under (Athar & Nasir, 2005).

Given the uses, there are also side effects of consuming the plant. When consumed orally, the tree tobacco is likely unsafe. This is because it contains a substance known as anabasine. This chemical is poisonous and may cause the heart to stop beating, severe muscle weaknesses, breathing problems, brain damage, high blood pressure, severe vomiting, seizures, spasms and in some instances even death. There is limited information regarding its safety on application to the skin.

The dosing of this plant depends on the user's age, health and other conditions. There is no scientific information regarding the dosage but it would be relevant to consult a pharmacist, a physician or a health care professional regarding the relevant direction of using the same. In the case of an individual who is pregnant or breastfeeding, nicotiana tobacco is likely to be unsafe because of the chemical anabasine contained in the product. For pregnant mothers, it may cause birthing defects.

Neem

The neem tree, also known as the margosa tree, is native to India. It is a small to a medium-sized evergreen tree that is cultivated in many tropical regions throughout the world. The scientific name of the plant is Azadirachta indica), also called nim or margosa, it is of the mahogany family (Meliaceae), valued as a medicinal plant and grows very fast. The plant is native to the Indians and the dry areas of South Asia. After centuries the plant was introduced to different parts of Africa, the Caribbean and numerous parts of south and central America. For a very long time, the plant has long been used in Ayurvedic and folk medicine. It has also been used for organic farming applications and cosmetic purposes.

Figure 59: Neem Tree.

Source: image by Leopard point Beach Resort and SPA

The plant is widely cultivated in tropical regions around the world because of its many beneficial properties. It is also grown for its attractive form and good-looking leaves, which are used to make medicine, insecticide, dye, biopesticide, food ingredient. farmers in India have generally appreciated the plant for a very long time for its multiple beneficial properties. It is especially useful to control minor crop infestations, improve general soil conditions, deter pests, and enhance overall crop health. When applied as directed, neem oil will not harm the farmer's crop or other plants in the area. Apart from the fact that it has insecticidal and pesticidal properties, the plant has fungicidal, anti-viral and anti-bacterial properties. There are hundreds of chemical components that are present in the plant that allow it to have a range of therapeutic benefits. The ingestion of the plant is known to reduce plaque formation if the gel extract is applied to the teeth and gum, be able to heal stomach and intestinal ulcers and also reduce the severity of psoriasis symptoms in individuals (Zhou et al., 2014).

For adults, neem is possibly safe when applied to the mouth or skin within the time limit recommended by the doctor. When taken in large doses or for long periods of time, the plant is known to harm the liver and kidneys.

The appropriate dosing of the neem plant is dependent on the user's health, age and several other conditions. Since there is limited knowledge regarding the dosage, it would be wise if the user consulted with a pharmacist, healthcare professional or a pharmacist.

Nicotine

Is scientifically described as nicotiana and can be in various species. Is a genus of the herbaceous plant. They are also in the form of shrubs. In most cases, they are referred to as tobacco plants. They are of the nightshade family. It is native of tropical America but have been neutralized in other parts of the world. In most cases, they are found in South Africa, New Zealand and Australia. One species is the Nicotiana rustica contains over fifty different alkaloids in concentrations varying between plants and slightly within a single plant, but containing at least nicotine and nornicotine. Most tobacco grown around the world today is Nicotiana tabacum, although this species also contains nicotine and other related compounds such as anabasine, anatabine, myosmine, and beta-amyrine.

This plant normally flowers and is used primarily for its psychoactive chemical content, which is toxic and considered an addictive drug. When nicotine is pure, that is, it is not diluted, it is said to be a very poisonous substance to the human body. There are those products that may contain pure nicotine but are non-tobacco. In most instances, these products contain tobacco-derived ingredients. In e-liquids, the amounts of nicotine can vary. Those with high levels of nicotine may be poisonous (Yi et al., 2016). As such, when using this product, it is important to note that they need not spill on your skin. The eyes would be very sensitive to such. One originally isolated product was "nicotianine", is a potent parasympathomimetic alkaloid widely used as an insecticide. It is widely used around the world to treat various diseases such as asthma, hay fever, allergies, and various skin problems and can be used to control weight. In other instances, they can be used as a stimulant.

The dosage of nicotine also depends on the user's health, age and several other conditions. The possible side effects of the product are similar to that of Nicotiana tabacum.

Nettle

Often referred to as the common nettle or stinging nettle. It is scientifically known as the *Urtica dioica.* It is a dioecious and perennial plant that to between 1 to 2 meters in the summer and it dies down during the winter period. The rhizomes and stolons are widely spread hence they tend to grow in clusters. The soft green leaves are between 3 to 15 centimeters long and are found opposite each other on an erect and green stem. It is considered to be native to Europe, large areas of temperate Asia and the northern part of western Africa. It is now quite common in many parts of the world especially if the area is temperate and there is moist soil.

When nettle comes in contact with the skin it produces an inflammatory effect on a person's skin and it is referred to as a stinging, burning sensation and its commonly called the contact urticaria. The burning sensation is both by impaling the skin through spicules which causes mechanical irritation and there are biochemical irritants i.e., acetylcholine, histamine and serotonin among other chemicals. As the old English store recorded the nettle is one of the nine plants invoked in the pagan Anglo-Saxon nine herbs charm in the tenth-century medicine (Van Jaarsveld, 2013). The stinger was believed to be a galactagogue which was a substance that promoted lactation. In some instances, urination was a process of flogging using nettle which in turn caused redness to provoke inflammation. This was considered to be a rubefacient and it was used to treat rheumatism.

Figure 60: the stinging nettle.

Source; image by unsplash

It also used in the treatment of diabetes and it is consumed by ingesting through the mouth and it is considered to reduce the blood sugar levels in individuals that have type two diabetes. It has continued to be used for hundreds of years to treat painful muscles and joints, anemia, gout, and arthritis. In recent people are using it to treat urinary problems in cases where there tends to be an enlarged prostate (Umadevi et al., 2012). The root is mostly used in Europe to treat Benign Prostatic Hyperplasia and it is also believed to be used in combination with other herbs to relieve symptoms such as post urination dripping, constant urges to urinate, and reduced urine flow and incomplete emptying of the bladder. The leaves and the stems are used to treat arthritis and also relieve sore muscles. It was also discovered in some studies that it could also reduce sneezing and itching in people that suffer from hay fever.

The hairs of the stinging nettle contain chemicals that help in relieving pain and they also have anti-inflammatory effects. This help in treating individuals with some form of allergy. When the nettle is consumed via ingestion it is recommended to be used for up to one year as it may cause diarrhea, stomach upset and even constipation. In some instances when it is applied to the skin it may be safe but touching the plant in itself may cause skin irritation which to some individuals may lead to hospitalization.

O

Ochna Serrulata

Several names have been used to describe this plant; these names include; small-leaved plane, mickey mouse plant or mickey mouse bush, bird's eye bush and also carnival Ochna. They are referred as mickey mouse bushes because of their ripe black fruit which when turned upside down resembles the cartoon character mickey mouse ears and also the bright-red sepals which in turn resemble his trousers. it is said to have originated from southern Africa and the people of South Africa grow it as an ornamental plant in their gardens but in Australia and New Zealand, it is seen as an invasive weed.

It is a deciduous plant or shrub and it is known to belong to the family Ochnaceae and it has been known to grow in the zones between nine to eleven. It grows to a height of between 4.00 to 8.00 feet and it spreads between 3.00 to 4.00 feet (Thakur, 2011). The bloom time for this shrub is between May and June and the flowers are yellow whereas the calyx is red. They can grow in the sun fully and sometimes in n partly shaded places or areas, they only require medium amounts of water and their maintenance is quite low. They can be used for several purposes which include but are not limited to being used as hedges because of their aesthetic value and they are also used to naturalize the compound and, in some cases, they can be used for medicinal purposes.

Figure 61: The *Ochna Serrulata.*

Source: weed identification Brisbane city council

They are found in different habitats which are forest margins, some forested areas in some grasslands and open rocky slopes. They have glossy green and they also have narrow elliptic leaves and their margins are serrated the flowers have a fragrance and they have a five-petaled flower that is yellow and the stamens are also yellow and they bloom from short lateral branches from spring to early summer. The yellow petals tend to fall fairly quickly and the floral receptacle the sepals appear to be swollen and they then turn bright red. By early summer they can be seen to have red flowers instead of red sepals. The fleshy fruits, appear pear-shaped and are about five to six per calyx and they mature to black in the early summer and at this point, they begin to resemble the shape of a mickey mouse. They are considered to be very invasive in Australia and New Zealand because these fruits attract a lot of birds and in this case, the birds tend to spread the seeds across wide areas through their fecal matter.

This plant was first described by a German naturalist Christian Ferdinand Friedrich Hochstetter and then it was given its scientific name by Wilhelm Gerhard Walpers. The specific part of the scientific name was derived from the Latin name *Serrula* which means little saw and this refers to the margins of the leaf which are fine-toothed or simply serrated (Stanisavljević et al., 2007).

The South African tribes used the roots of *O. serrulate* in the form of decoctions for the treatment of some diseases such as osseous illnesses and also in hemorrhoids. There are biological studies that have been carried out and there have been different biological activities of the different Ochna species and it has been reported that it is an analgesic and anti-inflammatory, anti-HIV, antibacterial and cytotoxic and also antimalarial. The *O. serrulate* has been known to have genotoxic properties.

In Brazil, they collect the stems and leaves in the city of San Jose in Santa Catarina. The air-dried leaves and stems are then triturated and macerated separately with 80% ethanol at room temperature for up to fifteen days. The solvent was then removed by the process of rotary evaporation and the extract was then suspended in methanol then it was refrigerated to decant the non-soluble material. The residue was then filtered and the solution was added to 30% ethanol then it was partitioned with organic solvents and they were of increasing polarity. This procedure eventually yielded three fractions i.e., hexane, ethyl acetate and butanol respectively.

In Indian traditional medicine, the Ochna was used in the treatment of different ailments such as ulcers, constipation, epilepsy, sores, cancer, lumbago, asthma, menstrual complaints and also in snake bites. The root bark was used as a digestive tonic whereas the boiled leaves were regarded as a cure for ulcers and lumbago. A decoction that was derived from the roots of these plants was used in the treatment of menstrual complaints and asthma. In some traditional African communities especially in the southern part of Africa most especially Mozambique, Zambia, Zimbabwe and the eastern part of South Africa they use the *O. serrulata* to treat diseases that attacked the bones.

Ocimum Sanctum

It is commonly referred to as holy basil and it is an aromatic and perennial plant that belongs to the family Lamiaceae. It is native to the Indian subcontinent and it is also widespread as a cultivated plant throughout the southeast Asian tropics. It is also sometimes referred to as tulsi. The most common synonyms that are used for this plant are; *Genuisporum tenuiflorum* (L) Merr, *Lumnitzera tenuiflorum* (L) Spreng, *Moschosma tenuiflorum* (L) Heynh, *Ocimum anisodorum* F. Muell, *Ocimum caryophyllinum* F. Muell, Ocimum hirsutum Benth, Ocimum inodorum Burm. F., Ocimum monahorum L., Ocimum tenuiflorum L., Ocimum scutellarioides Willd. Ex Benth, Ocimum subserratum B. Heyne ex Hook. F., Ocimum tomentosum Lam, Ocimum villosum Roxb. Nom. Illeg, Plectranthus monachorum (L) Spreng.

This plan is mostly used for religious and traditional medicine purposes and some individuals use it for their essential oils. It is widely used in most areas as herbal tea i.e. in Ayurveda and it is said to have a place within Vaishvana tradition in the Hindu religion. The variety that is mostly used by the Thailand people is referred to as Thai Holy Basil ad it varies from the Thai basil and the chi neang vorng.

Figure 62: The holy basil.

Source: Image by pharmacognosy journal

The plant is erect and has many branches and it grows to a height of about 30 to 60 centimeters and it has tall hairy stems. The leaves appear to be green or purple and their appearance is simple and is also petiole with a blade that is ovate and is approximately 5 cm long and the margin is slightly toothed. The leaves are also very strongly scented. The flowers of this plant are purplish and are placed in close whorls on elongated racemes.

The plant is said to have originated in north-central India. This came about after research was carried out in a large phylogeographical study of these species which they used the chloroplast genome sequences by a group of researchers from the Central University of Punjab.

This plant has several extracts that can be derived from it and the phytochemical constituents include, beta-caryophyllene, carvacrol, eugenol, oleanolic acid, linalool, rosmarinic acid and ursolic acid (Seth, 2003).

It is mostly used for religious purposes by the Hindus and also some Indians use it for the treatment of certain diseases. In some areas because of the tulsi's strong scent, they mix the dried leaves with some grains then it is used to repel the insects. In the religious aspect, the Hindus worshiped the plant as an avatar of Lakshami and they are planted in the courtyard of Hindu households or in the Hanuman temples. The ritual of lighting of the lamps in the evenings a ritual referred to as Kartik includes the worship of the tulsi plant.

On the medicinal arts of this plant, it is known to aid in individuals that have coughing problems, dysentery, eye diseases, arthritis, asthma, diarrhea, fever, gastric ailments, indigestion and many other several ailments. The plant is known to be beneficial the body and for the heart, the tulsi is known to contain eugenol which is a phenol compound and it is known to inhibit copper-dependent LDL oxidation and also to prevent or reduce the iron-mediated lipid peroxidation and eventually lead to a reduction in total cholesterol, phospholipids, and total lipids in major body organs i.e. the liver, kidneys and the heart. It is also essential in Alzheimer's and this is because the tulsi ameliorates hydrogen peroxide-induced neuronal damage through the use of its antioxidant defense mechanism and it may also be used to treat oxidative stress neuronal disorders. In the field of diabetes, the holy basil helps in the prevention of diabetes and diabetic complications through its antioxidant activity and this is carried out by reducing the plasma level of thiobarbituric acid reactive substance (TBARS) and they help maintain the antioxidant markers like glutathione peroxidase, dismutase and catalase. With all these the antidiabetic effect results to the lowering of blood glucose, glycosylated hemoglobin, lipid peroxide, and FFA which has been attributed to stimulating the body to secrete more insulin.

The holy basil is said to prevent anti-aging and this is carried out by the vitamins C and A which act as good antioxidants and in turn they protect the skin from almost all damages that are or may be caused by free radicals. They also help in the treatment of kidney stones as it acts as a mild diuretic and detoxifying agent which in turn helps in lowering the amount of uric acid in the body the acetic acid in the basil also helps in the breakdown of the stones. It is also known to prevent headaches as t also helps to relieve migraines. The holy basil also helps in killing bacteria and preventing infections (Saslis-Lagoudakis et al., 2014). The extract eugenol helps in fighting skin-related disorders, they, therefore, help the skin both internally and externally. It has been in use since the old world in the treatment of fever and it is one of the prime ingredients that are used in the formulation of ayurvedic medicines and the making of some home remedies. It is also used as a natural mouth freshener and also as a mouth disinfectant. It can also be used to treat mouth ulcers. They are responsible for destroying the bacteria that cause dental cavities, bad breath, plaque, and tartar and they also protect the teeth. They are also involved in the treatment and prevention of infection in the cardiovascular system. It also cures some diseases such as bronchitis and tuberculosis. It is known to be a rich source of vitamin K and it's known to play a vital role in bone and heart health.

Oryza Sativa

The common name is Asian rice and it is the plant that in English it is referred to as rice. It is the common type of farmed rice whose cultivars are very common worldwide and it was first domesticated in the Yangtze River basin in mainland China between 8,200 to 13,500 years ago. List *Oryza aristata Blanco, Oryza communissima Lour., Oryza denudata* (Desv.) Steud., *Oryza elongata* (Desv.) Steud., *Oryza formosana Masam.* & Suzuki, *Oryza glutinosa* Lour. *Oryza marginata* (Desv.) Steud. *Oryza montana* Lour. *Oryza mutica* Steud. *Oryza palustris* Salisb. *Oryza parviflora* P. Beauv. *Oryza perennis Moench Oryza plena* (Prain) N.P. Chowdhury Oryza praecox Lour. *Oryza pubescens* (Desv. Steud. *Oryza pumila* Steud. *Oryza repens* Buch. -Ham. ex Steud. *Oryza rubribarbis* (Desv.) Steud. *Oryza sativa subsp. indica* Shig.Kato *Oryza sativa subsp. japonica* Shig.Kato and Oryza segetalis Russell ex Steud. Are all the synonyms that are associated with *Oryza sativa*.

Figure 63: oryza sativa.

Source: image by Wikipedia.

Oryza sativa is a grass and it is known for being very easy to genetically modify and it is known by scientists as a model organism for cereal biology.

It is known to contain two major subspecies and they include the short-grained and sticky japonica or sinical varieties and the non-sticky long-grained varieties. The variety japonica is mostly cultivated in dry fields in temperate east Asia, the upland areas of Southeast Asia and the elevations of southeast Asia while the indica varieties are mainly grown submerged throughout the tropical parts of Asia. Rice is known to occur in various colors i.e. purple, white, brown, black and red rice. The black rice which is also referred to as purple rice is a range of various rice types some of which may be glutinous rice, the varieties can include Thai jasmine black rice and Indonesian black rice.

Until recently most people were looking at rice as a source of food but recent studies that have been carried out have shown that rice contains a lot of different properties and some of which can be seen in ancient literature. The ancient rice-growing communities in Asian countries such as Myanmar, China, Malaysia, Thailand, Pakistan and India have attributed some medicinal properties to rice in addition to its being a good source of starch. In these texts, it is seen that brown rice is mentioned as the perfect food. In China, the medicinal value of rice has been traced as far back as in the year 2,800 BC when it has been recorded as being used by royal Chinese physicians for healing. I was known to provide a sense of calm, tranquility and peace to those individuals that were easily upset. The dried rice that had just sprouted was used to aid in the process of digestion, tone muscles and also to expel gases from the body (Nazim et al., 2020. The Chinese also believe that rice helps in strengthening the spleen and also the stomach, it increases appetite and cures indigestion. In the traditional Malaysian society in the medical writings, they used prescribed boiled rice "greens" as an eye lotion and for use also in acute inflammation of inner body tissues. The application of dried powdered rice was recommended for use in skin ailments. In traditional Cambodian society, the hulls that were derived from mature plants were considered to be useful in the treatment of dysentery. Hulls of three-month-old rice were used as a diuretic. In the Philippines, rice polish or bran or "*tikitiki*" were extracted and used as an excellent source of vitamin B to cure and prevent beriberi.

Rice is also used as a diuretic a used to treat urinary dysfunctions. Excessive lactation, indigestion and poor appetite can be treated using the seeds from rice. On the other hand, the rhizome is taken internally and can be used to treat tuberculosis, chronic pneumonia and night sweats. The rhizomes can also be used after being harvested at the end of a panting season then dried and can then be used in making decoctions.

Olive Leaf

It has long been used in traditional medicine and the olive leaf extract is derived from the olive tree i.e. *Olea europaea* and it has a variety of health benefits under its name. research has shown that oleuropein which is the main component of the olive leaf extract has several benefits i.e. it acts as an antimicrobial, anti-inflammatory, immune-stimulating, antioxidant and antibacterial. In the past, the extract from the olive leaf has been used for the treatment and prevention of a variety of inflammations and infections.

Olive leaf extracts are known to aid in the control of diabetes. A study that was conducted on 79 patients that were known to have type 2 diabetes took a supplement of olive leaf extract or a placebo every day for 14 weeks. By the end of the study, it showed that participants that had taken the olive oil extract showed a significant decrease in their blood sugar levels as compared to individuals in the control group. In the treatment of hypertension, the use of olive leaf extract is showing some great potential. The research was conducted on 60 hypersensitive patients and some were given olive leaf extract and some were given a placebo. By the end of the test period which was six weeks it was found out that there is a significant decrease in blood pressure, LDL cholesterol, total cholesterol and inflammatory markers in individuals that had been given the olive leaf extract (Tuttolomondo et al., 2014). Olive leaf extract are also known to have antiviral properties and they are sometimes used in holistic medicine to treat some types of viruses. In the laboratory, it can be established that the extract could effectively fight mononucleosis, rotavirus and herpes and they can also be useful in the fight against influenza and the HIV virus.

Figure 64: Olive leaves.

Source: Yeni Safak

The extracts that are derived from olive leaf's come in various ways and these include capsules, soft gels and also as tinctures. There is no recommended dosage for the extract. The standard dosage is between 500 to 1000 mg daily. I most of the supplements it is recommended that the dose is split to 250 mg to 500 mg for four or two times a day respectively with food. There is little research that has been conducted on the side effects of olive leaf extract and it has been recorded sometimes that the extract can cause or trigger mild effects such as stomach pain and headaches (Teixidor-Toneu et al., 2018). There are also some individuals that may develop allergic reactions. The pollen that is derived from the olive leaf can also trigger severe respiratory complications. The olive leaf extract can be taken and interact well with other medicines but before combining the extracts with other medicine one is required to receive guidance from an expert i.e. a doctor or a physician. They can counteract blood pressure medications as it can cause the blood pressure to become too low. It can also cause hypoglycemia which is production of excess insulin I the body. Due to its antioxidant properties it can counter certain chemotherapy drugs.

Olive Oil

Olive oil is a major component of the Mediterranean diet. It is also very rich in antioxidants. The main fat that it contains is monounsaturated fatty acids (MUFAs) which many experts consider it to be a healthy fat. Olive oil comes from olives which are the fruits from the olive tree. The olives are traditionally from the Mediterranean region. Most individuals make oil by pressing the whole olives. They are mainly used in cooking, medicine, fuel, soaps, cosmetics and in traditional lamps. Many studies have been conducted on the health benefits of olive oil the extra virgin olive oil is rich in antioxidants and they help prevent cellular damage that may be caused by molecules called free radicals. The antioxidants are known to neutralize the free radicals. If the body has too many free radicals, they may cause oxidative stress and this in turn leads to cell damage and may lead to development of diseases such as cancer.

In the Mediterranean diet olive oil is a very essential component and it has been noticed that people who consume this diet have a high life expectancy and this includes a lower chance of dying from cardiovascular disease as compared to individuals who consume other diets. In a 2018 review that was done by the Food and Drug Administration (FDA) and the European Food Safety Authority recommended that consuming two table spoonsful of extra virgin oil a day will reduce the risk of cardiovascular disease and inflammation. The polyphenols that are present in extra virgin oil may offer protection from brain dysfunction, cancer, atherosclerosis, stroke and cardiovascular disease. The polyphenols are a type of antioxidant. It has been recommended that the extracts that are present in extra virgin olive oil are able to help in protecting the nervous system and they could be useful in treatment of depression and anxiety.

The metabolic syndrome which is a condition that is characterized by a combination of risk factors that may lead to risk of contracting diseases which include high blood pressure, high blood sugar levels and obesity. It is believed that extra virgin olive oil has extracts that help in reducing metabolic syndrome.in some other studies they show that substances that are present in olive oil help prevent individuals from getting breast cancer (Silva et al., 2010). These extracts are known to prevent oxidative damage, inflammation and epigenetic changes which provides high risk for contracting cancer. There are some scientists in 2016 who suggested that including extra virgin olive oil in the diet helps in the prevention of Alzheimer's disease and this may be due to the fact that it protects the impact on blood vessels in the brain. Oleocanthal is a phenolic compound and is an extract that is readily found in extra virgin olive oil and it could help stop or slow the progression of Alzheimer's disease.

Onions

They are thought to have originated from central Asia in modern-day Iran and Pakistan. It is thought that prehistoric people ate onions even before it was farmed or domesticated.

Phytochemicals and/or phytonutrients are compounds that naturally occur in fruits and vegetables and they are capable of reacting together with the human body to trigger healthy reactions. Flavonoids are responsible for the color pigments found in fruits and vegetables and studies have shown that they may be responsible in helping reduce the risk of cardiovascular disease, stroke and even Parkinson's disease. The particular flavonoid that is present in onions is quercetin which acts as an antioxidant and may be linked to the prevention of cancer. This extract also has a host of other benefits which include reducing symptoms of bladder infections, promoting of prostate health and helping in lowering blood pressure. The other extracts that can also be found in onions include disulfides, cepaene, trisulfides and vinyl dithiins. They are very useful to the body and they contain anticancer and antimicrobial properties. They are also a good source of antioxidants and this characteristic gives it the sweetness and aroma. Onions are healthy whether they are cooked or eaten raw although the raw onions tend to contain more or higher levels of organic sulfur. The outer parts of the onions are regarded to contain a higher concentration of flavonoids hence it is recommended that one should peel as little as possible in order to preserve as many nutrients as possible.

Figure 65: The common bulb onion.

Source: image by www.innovationhomecare.com

Although the risks are not that serious eating onions have been known to cause problems for some individuals. The carbohydrates that are present in onions can cause gas and bloat. Onions also when consumed raw are known to worsen heartburn in individuals that may suffer from gastric reflux disease or chronic heartburn. Green onions are known to contain a lot of vitamin K and hence they may interfere with blood-thinning drugs and hence which decreases blood thinner function. Here are cases of food intolerance o onions but this is very rare and for an individual with onion allergies, they tend to experience the following i.e. rashes and red, itchy eyes.

Oregano

This is a mint that belongs to the Lamiaceae family. It has been used for many years to flavor dishes and to treat some health conditions. It also features in the Mediterranean diet. The romans and reeks in the old days associated it with joy and happiness. There are different types of oregano and the most common type is *Oregano vulgare* and it is commonly referred to as Spanish thyme or the wild marjoram. Individuals mostly use the oil as a supplement or as an aromatic oil (Sepúlveda-Cuellar et al., 2021). It contains extracts such as thymol, ocimene, caryophyllene which give the herb the scent and the flavor, carvacrol, limonene and treponeme which are antioxidants. They have been used as medicine for a long and they are used to treat ailments such as colds, cramping, diarrhea, indigestion, aching muscles, asthma and skin sores. They also act as antibacterial agents.

Although oregano is very beneficial to the body it is recommended that one should avoid it for two weeks in case one is scheduled for surgery as it increases the risk of bleeding. It may sometimes cause skin irritation if used in large quantities especially if used in a bath. It also tends to reduce the concentration of body sugar and it also reduces the ability of the body to absorb iron, zinc and copper.

Figure 66: Ooregano.

Source: Image by the spruce

P

Papaver Somniferum

Synonyms

Papaver somniferum, sometimes known as the opium poppy or bread seed poppy, is a flowering plant that belongs to the Papaveraceae family. It is the species of plant that yields both opium and poppy seeds, as well as being a desirable aesthetic plant in gardens. Its native area is most likely the eastern Mediterranean, but it has been naturalized across most of Europe and Asia as a result of ancient importation and cultivation.

Usage

Poppy extracts have long been used to reduce smooth muscle tension, making them potentially effective in the treatment of diarrhea and stomach cramps. As a sedative, analgesic, and antitussive, the extract has been utilized. Poppy seed oil is a carrier for delivering chemotherapy and diagnosing fistulae. However, no clinical trials have been conducted to back up these claims. The opium poppy is used to make morphine.

Extracts

Morphine alkaloids have a wide range of pharmacologic actions. Codeine and morphine are sedative analgesics that calm smooth muscular tone, making them effective for treating diarrhea and stomach cramps. Antitussives include codeine and its derivatives. Papaverine enhances cerebral blood flow by relaxing involuntary smooth muscles. Different receptor-dependent functions can be triggered by modifying the alkaloids chemically. For millennia, the addictive properties of opium alkaloids have been known (Salim et al., 2019).

Dosage

35-250 grams of poppy seed are mixed with a drink or yogurt and consumed by mouth to diagnose an improper connection between the colon and the bladder. Urine is then monitored for the next 48 hours.

Side Effects

Morphine alkaloids have a wide range of pharmacologic actions. Codeine and morphine are sedative analgesics that calm smooth muscular tone, making them effective for treating diarrhea and stomach cramps. Antitussives include codeine and its derivatives. Papaverine enhances cerebral blood flow by relaxing involuntary smooth muscles. Different receptor-dependent functions can be triggered by modifying the alkaloids chemically. For millennia, the addictive properties of opium alkaloids have been known.

Persea Americana

Synonyms

The avocado tree (*Persea americana*), which is said to have originated in south-central Mexico, belongs to the Lauraceae flowering plant family. The plant's fruit, often known as an avocado (or avocado pear or alligator pear), is a huge berry with a single enormous seed.

Uses

The herb has a wide range of medical applications in general. Leaf extracts have been proven to have antihypertensive and anticancer properties. Dysentery, coughs, high blood pressure, liver issues, and gout are all treated using the leaves. The fruits are used to decrease blood cholesterol, encourage hair growth, soothe and heal skin issues, and as an aphrodisiac. Crushed seeds are used to alleviate a variety of skin problems. Seeds also produce oil, which is used in cosmetics, and dye, which is used to mark clothing. Although the wood is relatively soft, it is quite weighty. Because it is brittle and not long-lasting, it is only suitable for light construction.

Extracts

- Leaf extracts have shown antihypertensive activity
- The oil from the seeds contains steroids that are used for pharmaceuticals
- The seed extract has an erythroagglutinating property

Side Effects

When eaten in meal amounts, avocado is generally safe and effective for most individuals when consumed by mouth. Keep in mind that avocados are high in calories due to their fat content. There is insufficient trustworthy evidence about the safety of avocado as a medicine when taken by mouth (Schultz et al., 2020).

When applied to the skin, avocado is likely safe to use for up to three months. Although one person who used a special avocado oil plus vitamin B12 cream for psoriasis reported minor itching, it normally has little negative effects.

Dosage

Depending on the number of other fats and calories in the diet, the amount of avocado used varies. In rare cases, avocados have been substituted for other fats by eating 0.5 to 2 avocados per day.

Picrorrhiza Kurroa

Synonyms

Picrorhiza is a Himalayan plant that thrives in the himalayas. The root and rhizome (underground stem) are used for healing by people, particularly Ayurvedic practitioners. Picrorhiza has been heavily exploited to the point of extinction.

Uses

- A skin condition in which white patches appear on the skin (vitiligo). Adults and children with vitiligo appear to benefit from taking picrorhiza by mouth for up to a year in combination with a medicine called methoxsalen, which is taken by mouth and applied to the skin.
- Asthma. Taking picrorhiza orally for up to 12 weeks appears to not affect asthma symptoms or lung function.
- Inflammation (swelling) of the liver (hepatitis). Early research suggests that taking picrorhiza by mouth for two weeks can help persons with acute viral hepatitis reduce symptoms like anorexia, nausea, and overall discomfort.
- A lung illness that makes breathing difficult (chronic obstructive pulmonary disease or COPD).
- Rheumatoid arthritis is a type of arthritis that affects the joints (RA).

Side Effects

When eaten by mouth for up to a year, Picrorhiza is considered safe for most adults. In certain people, it can induce vomiting, rash, anorexia, diarrhea, and itching.

Dosage

Vitiligo is a skin condition that causes white patches to appear on the skin. Picrorhiza rhizome powder, 200 mg twice a day, in combination with methoxsalen, a drug taken orally and applied to the affected skin.

Extracts

Indiscriminate extraction of *Picrorhiza kurrooa* is a serious threat to the population of this Himalayan medicinal plant.

Phaseolus Calcaratus

Synonyms

The warm-season annual vine legume *Vigna umbellata*, formerly *Phaseolus calcaratus*, produces yellow blooms and little edible beans. It's also known as ricebean or white kidney bean. It is a little-known, little-researched, and little-exploited resource at the moment. It is a minor food and fodder crop that is commonly produced as an intercrop or mixed crop with maize (*Zea mays*), sorghum (*Sorghum bicolor*), or cowpea (*Vigna unguiculata*), as well as a solitary crop in the uplands on a small scale. Rice bran, like the other Asiatic Vigna species, is a warm-season annual with a limited life span (Asnake et al., 2016).

Figure 67: Rice bean (Vigna umbellata) before and after harvesting.

Source: https://www.feedipedia.org/node/234

Uses

- Obesity. Many early studies demonstrate that Phaseolus vulgaris can assist overweight persons to lose weight and shrinking their waistlines.

- Colon cancer and rectal cancer are two types of cancer. Eating Phaseolus vulgaris beans, such as baked beans, kidney beans, pinto beans, lima beans, and navy beans, has been related to a lower risk of recurrence in persons with advanced colon and rectal cancer, according to preliminary studies.
- Diabetes. Early research reveals that consuming a three-times-daily dose of Phaseolus vulgaris bean pod, white mulberry, and bilberry for two months could reduce blood sugar levels by about 25% in persons with type 2 diabetes.
- Infections of the kidney, bladder, or urethra (urinary tract infections or UTIs).
- Kidney stones.
- Other conditions.

Dosage

The white kidney bean variation of *Phaseolus vulgaris* has been the subject of the majority of studies. White kidney bean extract doses of 800-1000 mg were taken two or three times day for up to 12 weeks. Lower doses of white kidney bean extract, ranging from 445 to 1000 mg daily for up to 13 weeks, have also been used to reduce body fat, but these lower doses do not appear to reduce body weight.

Side Effect

There is some indication that *Phaseolus vulgaris* bean extracts, when taken by mouth for a short period, are POSSIBLY SAFE for most individuals. The majority of the study has focused on a white kidney bean extract from the *Phaseolus vulgaris* species (Phase 2, Pharmachem Labs). In doses of up to 3 grams per day for 2-3 months, this substance looks to be safe (Saslis-Lagoudakis et al., 2011). Other white kidney bean extracts, when administered for up to three months, appear to be safe. Eating a lot of fresh *Phaseolus vulgaris* bean husks could be dangerous. Lectins are compounds found in raw husks. Stomach upset, nausea, vomiting, and diarrhea are all possible side effects of these drugs. These compounds are destroyed by cooking.

Piper Nigrum

Synonyms

Black Peppercorn, Extrait de Poivre, Grain de Poivre, Hu Jiao, Kali Mirchi, Kosho, Marich, Maricha, Miris, Peber, Peper, Pepe, Peppar, Pepper, Pepper Extract, Peppercorn, Pfeffer, Pimenta, Pimienta, Pimienta Negra, Pipar, Piper, Piper nigrum, Piperine, Pippuri, Poivre, Poivre Noir, Poivrier, Schwarzer Pfeffer, Vellaja.

Figure 68: Black pepper fruits are the source of one of the world's most widely and frequently used spices. Black, white and green peppercorns all come from Piper nigrum and are popularly used as a hot and pungent spice for flavoring food. Black pepper is also used in traditional medicine, particularly for digestive ailments.

Source: http://plantsoftheworldonline.org/taxon/urn:lsid:ipni.org:names:682369-1

Uses.

- Arthritis.
- Asthma.
- Bronchitis.
- Cancer.
- Cholera

- Colic.
- Depression.
- Diarrhea..
- Skin that is discolored (vitiligo).
- Dizziness.
- Gas.
- Headache.
- Mites create itchy skin (scabies).
- Measles.
- Menstrual cramps
- Pain in the nerves.
- Pain.
- Sex Drive
- Nose congestion.
- Infection of the sinuses.
- Stomach ache.
- Loss of weight.
- Other circumstances.

Side Effects

When eaten by mouth in proportions usually seen in foods, black pepper is harmless.

When used as medicine, black pepper is safe when taken by mouth and when applied to the skin as an oil. Black pepper oil is usually free of adverse effects. Black pepper may have a searing aftertaste. It has been claimed that taking excessive doses of black pepper by mouth, which can mistakenly enter the lungs, can result in death. This is especially true when it comes to kids.

Dosage

The right amount of black pepper to take depends on a variety of factors, including the user's age, health, and other circumstances. There is currently insufficient scientific evidence to define an appropriate dose range for black pepper in children and adults. Keep in mind that natural products aren't always safe, and that dosages are crucial. Before using, be sure to read the product label and consult your pharmacist, physician, or other healthcare expert (Reinaldo et al., 2020).

Psidium Guajava

Synonyms

The common guava, also known as yellow guava or lemon guava, is a tiny evergreen shrub or tree native to the Caribbean, Central America, and South America. It is easily pollinated by insects; when farmed, it is mostly pollinated by Apis mellifera, the common honey bee.

Uses

- Colic.
- Ulcerative colitis.
- Vomiting.
- Gas (flatulence).
- Diabetes.
- Cough.
- Cataracts.
- High cholesterol.
- Heart disease.
- Cancer.
- Obesity.
- Pain.
- Hair growth.
- Skin and nail health.
- Sleep.
- Skin infections , when applied to the skin.
- Fever, when applied to the skin.
- Vaginal infections and irritation, when applied to the skin.
- Skin wounds, when applied to the skin.

Figure 69: Guava flower in Kabwe, Zambia.

Source: https://commons.wikimedia.org/wiki/File:Psidium_guajava.jpg

Side Effects

Guava fruit is safe to eat as food when ingested by mouth. When administered as a medicine for a short period, guava fruit and guava leaf extract are safe. Some people may experience temporary nausea or stomach pain as a result of guava leaf extract.

When applied to the skin or used as a mouth rinse, the guava leaf extract is safe. Some people may get skin discomfort as a result of it.

Dosage

The right amount of guava varies on a variety of factors, including the user's age, health, and other circumstances. There is currently insufficient scientific evidence to define an appropriate dosing range for guava. Keep in mind that natural products aren't always safe and that dosages are crucial.

Extracts

Because of its warming and invigorating properties, black pepper essential oil is widely used in the treatment of respiratory infections and the relief of muscular pains. This function is carried out by the pungent bioactive piperine, which necessitates the use of advanced technologies for a successful extraction (Rawat et al., 2020).

Punica Granatum

Synonyms

The pomegranate (*Punica granatum*) is a fruit-bearing deciduous shrub in the family Lythraceae, subfamily Punicoideae, that grows between 5 and 10 m (16 and 33 ft) tall.

Uses

- High blood pressure.
- A lung disease that makes it harder to breathe (chronic obstructive pulmonary disease or COPD).
- High levels of cholesterol or other fats (lipids) in the blood (hyperlipidemia).
- Hardening of the arteries (atherosclerosis
- Athletic performance.
- Heart disease.
- Tooth plaque.
- Diabetes.
- Serious kidney disease (end-stage renal disease or ESRD).
- Erectile dysfunction (ED).
- Muscle soreness caused by exercise.
- Symptoms of menopause.
- Muscle strength.
- Obesity.
- Thrush
- A serious gum infection (periodontitis).
- Prostate cancer.
- Rheumatoid arthritis (RA).
- Sunburn.
- A sexually transmitted infection caused by Trichomonas vaginalis

(trichomoniasis). Diarrhea.
- Hemorrhoids.
- Intestinal worm infestations.
- A hormonal disorder that causes enlarged ovaries with cysts (polycystic ovary syndrome or PCOS).
- Sore throat.
- Other conditions.

Extracts and Side Effects

When consumed by mouth, pomegranate juice is safe for the majority of people. The majority of people do not encounter any negative side effects. Pomegranate fruit can cause allergy responses in certain people. When eaten by mouth or applied to the skin, pomegranate extract is safe. Pomegranate extract has caused sensitivity in some persons. Itching, swelling, a runny nose, and difficulty breathing are all signs of sensitivity.

When consumed by mouth in significant amounts, the root, stem, or peel of the pomegranate is safe. Poisons are found in the root, stem, and peels.

When it comes to using pomegranate extract to the skin, it is safe.

Pyrrolizidine Alkaloids

Pyrrolizidine alkaloids (PAs), also known as necine bases, are a class of naturally occurring alkaloids based on the pyrrolizidine structure. Plants manufacture pyrrolizidine alkaloids as a defensive strategy against insect herbivores. In over 6,000 plants, more than 660 PAs and PA N-oxides have been discovered, with about half of them causing hepatotoxicity (Peng et al., 2017). They're most common in plants in the Boraginaceae, Asteraceae, Orchidaceae, and Fabaceae families; they're less common in the Convolvulaceae and Poaceae families, and they're found in at least one Lamiaceae species. Pyrolizidine alkaloids are found in 3% of the world's flowering plants, according to estimates. Honey can contain pyrrolizidine alkaloids, as can grains, milk, offal and eggs.

Unsaturated pyrrolizidine alkaloids are hepatotoxic, meaning they harm the liver. PAs are also responsible for hepatic veno-occlusive disease and liver cancer. PAs are carcinogenic. Pyrolizidine alkaloidosis is a disease caused by the ingestion of PAs.

The health risk linked with the use of therapeutic plants containing PAs, particularly borage leaf, comfrey, and coltsfoot in the West, and various Chinese medicinal herbs, is cause for concern.

When fed modest amounts of plants containing pyrrolizidine alkaloids, some ruminant animals, such as cattle, show no change in liver enzyme activity or clinical indications of poisoning.

Platelet-activating Factor

PAF, PAF-acether, or AGEPC (acetyl-glyceryl-ether-phosphorylcholine) is a powerful phospholipid activator and mediator of several leukocyte functions, including platelet aggregation and degranulation, inflammation, and anaphylaxis. It's also implicated in vascular permeability alterations, the oxidative burst, leukocyte chemotaxis, and phagocyte arachidonic acid metabolism amplification.

PAF is produced by a number of cells, including platelets, endothelial cells, neutrophils, monocytes, and macrophages, which are all involved in host defense.

PAF is continuously produced by these cells but in low quantities and production is controlled by the activity of PAF acetylhydrolases. It is produced in larger quantities by inflammatory cells in response to specific stimuli.

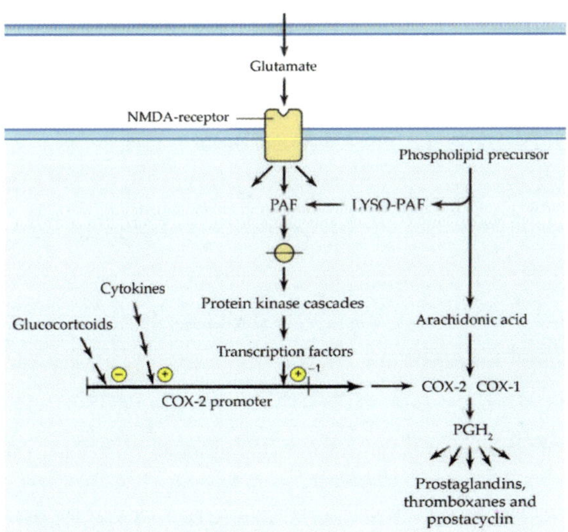

Figure 70: N-methyl-D-aspartate (NMDA) receptor activation by glutamate leads to phospholipase A2 (PLA2) activation and the generation of platelet-

activating factor (PAF) and of arachidonic acid. PAF is synthesized through other metabolic routes as well [119], and elicits its actions through a PAF receptor. PAF activates COX-2 gene expression through protein kinase cascades and transcription factors. The COX-2 promoter is also a target for cytokines (activation) and glucocorticoids (inhibition). COX-2 protein then catalyzes the conversion of arachidonic acid into prostaglandin H2 (PGH2), the precursor of eicosanoids. Constitutive activity of COX-1 also catalyzes this metabolic step.

Source: https://www.sciencedirect.com/topics/neuroscience/platelet-activating-factor

PAF is a hormone, cytokine, and other signaling molecule that is utilized to convey signals between nearby cells. The PAF signaling pathway can initiate inflammatory and thrombotic cascades, intensify them when other mediators are present, and mediate molecular and cellular contacts (cross talk) between inflammation and thrombosis (Ningthoujamet et al., 2012). Pathological inflammation can be caused by unregulated PAF signaling, which has been linked to sepsis, shock, and traumatic damage. PAF can be utilized as a local signaling molecule that travels short distances, or it can be distributed throughout the body and function through the endocrine system.

Parenteral

Parenteral delivery is defined by the US Food and Drug Administration (FDA) as drug administration by injection, infusion, and implantation or by some other route other than the alimentary canal.

Phenethyl Isothiocyanate (PEITC)

PEITC is a naturally occurring isothiocyanate whose precursor, gluconasturtiin, can be found in various cruciferous plants, particularly watercress.

PEITC has been investigated for its potential in cancer chemoprevention, particularly prostate cancer.

PEITC is created by the enzyme myrosinase from gluconasturtiin in terms of biosynthesis.

The chemical group –N=C=S is generated when the oxygen in the isocyanate group is replaced with a sulfur. Enzymatic conversion of metabolites called glucosinolates produces several natural isothiocyanates from plants.

Prostaglandins

Prostaglandins (PGs) are a class of physiologically active lipid molecules known as eicosanoids that have a wide range of hormone-like activities in animals. In humans and other animals, prostaglandins can be found in practically every tissue. Enzymatically, they are made from the fatty acid arachidonic acid. There are 20 carbon atoms in each prostaglandin, including a 5-carbon ring. They belong to the prostanoid family of fatty acid derivatives and are a subclass of eicosanoids.

The varied biological functions of prostaglandins are due to structural variations. In rare circumstances, a given prostaglandin might have diverse and even opposing actions in various tissues. The type of receptor to which a prostaglandin binds determines its ability to promote a reaction in one tissue while inhibiting the same reaction in another tissue (Nadda et al., 2020). They function as autocrine or paracrine factors in the presence of their target cells near the site of secretion. Prostaglandins vary from endocrine hormones in that they are created throughout the human body rather than in a single location.

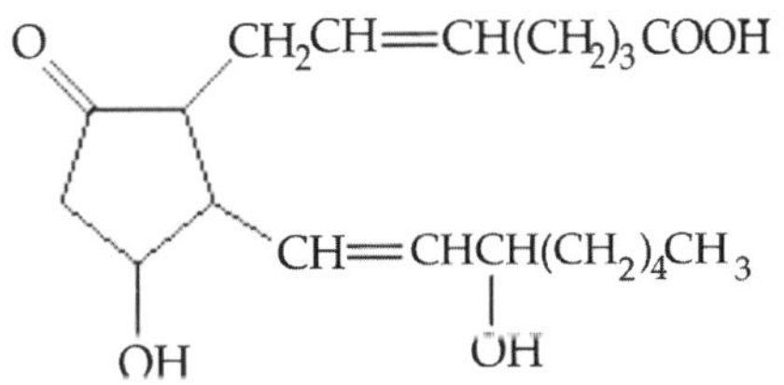

PGE_2 (a prostaglandin)

Figure 71: The prostaglandins are made up of unsaturated fatty acids that contain a cyclopentane (5-carbon) ring and are derived from the 20-carbon, straight-chain, polyunsaturated fatty acid precursor arachidonic acid.

Source: https://www.britannica.com/science/prostaglandin

Prostaglandins are potent, locally acting vasodilators that prevent blood platelets from clumping together. Prostaglandins play a function in inflammation through their role in vasodilation. They are produced in the walls of blood arteries and have the physiological purpose of preventing unnecessary clotting as well as controlling smooth muscle contraction. Thrombin (made by platelet cells) on the other hand, is a vasoconstrictor that aids platelet aggregation. Their name refers to their role in the production of clots (thrombosis).

Protein Kinases

Protein kinases, as opposed to kinases that modify lipids, carbohydrates, or other substances, preferentially modify other proteins by covalently adding phosphates to them (phosphorylation). By modifying enzyme activity, cellular location, or connection with other proteins, phosphorylation tends to result in a functional modification of the target protein (substrate). About 500 protein kinase genes are found in the human genome, accounting for about 2% of all human genes.

Protein kinases are divided into two categories. The majority of them are serine/threonine kinases, which phosphorylate the hydroxyl groups of serines and threonines in their targets, and the rest are tyrosine kinases, however there are others. Bacteria and plants both have protein kinases. Kinase activity can modify up to 30% of all human proteins, and kinases are known to influence the majority of cellular pathways, particularly those involved in signal transduction.

Premenstrual Syndrome (PMS)

PMS is a set of physical and mental symptoms that many women experience after ovulation and before their period begins. PMS is thought to occur in the days following ovulation because estrogen and progesterone levels drop rapidly if you are not pregnant. PMS symptoms fade away a few days after a woman's period begins, when her hormone levels begin to rise again (Miara et al., 2018).

Some women get their periods without or with just minor symptoms of PMS. Others may experience PMS symptoms that make it difficult to do daily tasks such as going to work or school. PMS symptoms that are severe could be a sign of premenstrual dysphoric disorder (PMDD). Researchers do not know exactly what causes PMS. Changes in hormone levels during the menstrual cycle may play a role. These changing hormone levels may affect some women more than others.

Physical symptoms of PMS can include:

- Swollen or tender breasts
- Constipation or diarrhea
- Bloating or a gassy feeling
- Cramping
- Headache or backache

- Clumsiness
- Lower tolerance for noise or light

Emotional or mental symptoms of PMS include:

- Irritability or hostile behavior
- Feeling tired
- Sleep problems (sleeping too much or too little)
- Appetite changes or food cravings
- Trouble with concentration or memory
- Tension or anxiety
- Depression, feelings of sadness, or crying spells
- Mood swings
- Less interest in sex

Parts Per Million

Similar term(s): ppm.

A weight to weight ratio used to describe concentrations. Parts per million (ppm) is the number of units of mass of a contaminant per million units of total mass.

Prostate-specific Antigen

PSA, also known as gamma-seminoprotein or kallikrein-3 (KLK3), P-30 antigen, is a glycoprotein enzyme that is encoded by the KLK3 gene in humans. PSA is a peptidase that is released by the epithelial cells of the prostate gland and belongs to the kallikrein-related peptidase family.

PSA is produced in the ejaculate to let sperm move easily by liquefying semen in the seminal coagulum.

It is also thought to aid in the dissolution of cervical mucus, allowing sperm to enter the uterus.

Medical uses

Clinical practice guidelines for prostate cancer screening vary and are contentious, in part because it's unclear if the advantages of screening outweigh the dangers of overdiagnosis and treatment. The PSA test has been approved by the Food and Drug Administration (FDA) in the United States for annual prostate cancer screening in males aged 50 and over. [a medical citation is required] Prior to administering the test, the patient must be advised of the risks and benefits of PSA testing (Menendez-Baceta et al., 2015).

As of 2018, the National Health Service in the United Kingdom neither mandates nor recommends PSA testing, but rather allows individuals to make their own decisions based on their doctor's opinion.

Protein Tyrosine Kinase

A phosphate group from ATP is transferred to the tyrosine residues of particular proteins inside a cell by a tyrosine kinase enzyme. Many cellular functions use it as a "on" or "off" switch.

Tyrosine kinases are part of the wider protein kinase family, which also attach phosphates to other amino acids including serine and threonine. Kinases phosphorylate proteins, which is an important method for conveying signals inside a cell (signal transduction) and controlling cellular activity like cell division.

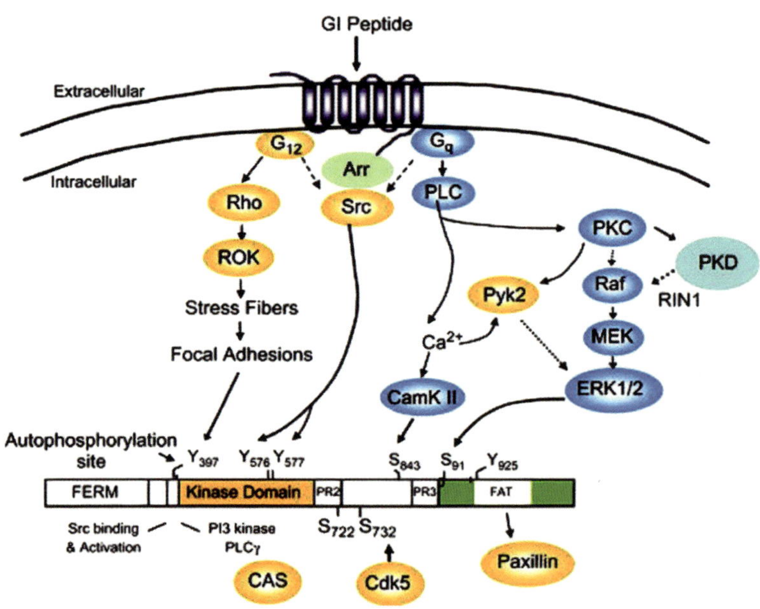

Figure 72: Signaling through the G12 subfamily (comprising Gα12 and Gα13) transduces GPCR signals into Rho activation, actin remodeling, assembly of focal adhesions, and tyrosine phosphorylation of the focal adhesion-associated proteins FAK, CAS, and paxillin and complex formation between FAK and Src. Notice that FAK integrates multiple signal transduction pathways by multi-site

tyrosine (e.g., Tyr397, Tyr576/577, Tyr861, and Tyr925) and serine (Ser722, Ser843, Ser910) phosphorylation. These pathways are implicated in both cell proliferation and cell migration. See text for details.

Source: https://www.sciencedirect.com/topics/biochemistry-genetics-and-molecular-biology/non-receptor-tyrosine-kinase

Protein kinases can get altered and locked in the "on" position, causing uncontrolled cell growth, which is an essential step in cancer development. As a result, kinase inhibitors like imatinib are frequently used to treat cancer.

The phosphate group is removed by a protein called tyrosine phosphatase, which is found in most tyrosine kinases.

Q

Quisqualis Indica

Synonyms

Combretum indicum, also known as the Rangoon creeper or Niyog-niyogan, is a vine with red flower clusters which is native to tropical Asia.

Use and Dosage

The herb is frequently utilized in traditional medicine, with a special emphasis on parasite removal. The entire plant is utilized.

The fruits are anthelmintic and are used to treat ascarids in particular. They produce nausea, vomiting, hiccough, and even unconsciousness in big dosages. The fruit's decoction is used as a vermifuge. The fruits are used to treat nephrotic syndrome.

A toothache-relieving gargle is made from a strong infusion of the fruit.

Side Effects

When properly prepared and consumed in moderation or as a small amount included in daily diet, Niyog-niyog fruit is safe.

However, when taken more than the commended dosage such as fruit/nuts are eaten on consecutive days or when fresh nuts are eaten. This may result to overdose and may have the following adverse effects:

- hiccups
- diarrhea, abdominal pain, distention
- nausea and vomiting
- unconsciousness

Quebracho

Synonyms

Quebracho is a common name in Spanish to describe very hard (density 0.9–1.3) wood tree species. The etymology of the name is derived from quiebrahacha, or quebrar hacha, meaning "axe-breaker".

Uses

- Asthma.
- Lung disorders.
- Cough.
- High blood pressure.
- Spasms.
- Fluid retention.
- Menstrual cramps.
- Fever.
- Increasing sex drive.
- Other conditions.

The bark is utilized in traditional medicine.

Figure 73: Quebracho colorado (Schinopsis balansae) wood.

Source: https://en.wikipedia.org/wiki/Quebracho_tree#/media/File:Quebracho_colorado_wood.jpg

It's important to distinguish between quebracho blanco (white quebracho) and quebracho colorado (colored quebracho) (red quebracho). They're both called quebracho, but they're made up of different substances. White quebracho is the subject of this information.

Asthma and lower respiratory tract problems are treated with quebracho. It's also used to relieve chest congestion and as a stimulant for the respiratory system (Muchugi et al., 2006).

High blood pressure, spasms, fluid retention, fever, and pain are some of the conditions quebracho is used to treat. It's also used to boost sexual desire.

Quebracho is used as a flavor in meals and beverages.

Side Effects

In food quantities, Quebracho is safe. However, there isn't enough data to say whether quebracho is safe in therapeutic doses. Drooling, headaches, sweating, dizziness, stupor, and tiredness are all possible adverse effects. It can produce nausea and vomiting in large dosages.

Dosage

The right dose of quebracho is determined by a variety of factors, including the user's age, health, and other circumstances. There is currently insufficient scientific data to define an appropriate dosing range for quebracho. Keep in mind that natural products aren't always safe, and that dosages are crucial. Before using, be sure to read the product label and consult your pharmacist, physician, or another healthcare expert.

Queen Anne's Lace

Synonyms

Beesnest Plant, Bird's Nest Root, Carotte Commune, Carotte Sauvage, Dauce Carotte, Daucus, Daucus carota, Garijara, Nan He Shi, Nid d'Oiseau, Queen Anne's Lace, Shikha-Mula, Zanahoria Silvestre.

Daucus carota, also known as wild carrot, bird's nest, bishop's lace, and Queen Anne's lace (in North America), is a white flowering plant in the Apiaceae family that is native to temperate Europe and southwest Asia but has naturalized in North America and Australia (Maruca et al., 2019).

Domesticated carrots are cultivars of the *Daucus carota subsp. sativus*

Uses

- Kidney stones and other kidney problems.
- Bladder problems.
- Gout.
- Diarrhea.
- Indigestion.
- Gas.
- Worm infestations.
- Pain in the uterus.
- Heart disease.
- Cancer.
- Water retention.
- Use as a nerve tonic.
- Use as an aphrodisiac.

- Starting menstruation (periods).
- Other conditions.

Side Effects

When taken by mouth in the proportions used in pharmaceuticals, wild carrot seed oil appears to be safe for most individuals. There isn't enough information to say whether wild carrot's above-ground sections are safe.

Wild carrot oil in high dosages might harm the kidneys and create nerve disorders. When exposed to the sun, wild carrots can create a rash and increase the risk of sunburn.

Dosage

The correct dose of wild carrot is determined by several factors, including the user's age, health, and other circumstances. There is currently insufficient scientific data to define an appropriate dosing range for wild carrots. Keep in mind that natural products aren't always safe and that dosages are crucial. Before using, be sure to read the product label and consult your pharmacist, physician, or another healthcare expert.

Extracts

The ethnobotanical uses of this species include applications in the treatment of cough, diarrhea, dysentery, cancer, malaria and tumors, and as an antiseptic, abortifacient, aphrodisiac, carminative, stimulant, stomachic and tonic. The major constituents isolated from the methanol extract of D. carota seeds by reversed-phase preparative high-performance chromatography were luteolin, luteolin 3'-O-beta-D-glucopyranoside and luteolin 4'-O-beta-D-glucopyranoside, three flavones (Martínez & Luján, 2011).

Queen'S Delight

Synonyms

Cockup Hat, Marcory, Queen's Root, Racine Royale, Raíz de la Reina, Silver Leaf, Stillingia, Stillingia sylvatica, Stillingia tenuis, Yaw Root.

The Queen's Delight is a type of plant. The root is utilized in traditional medicine.

People use queen's pleasure to cure liver disease, gallbladder issues, skin diseases, constipation, bronchitis, and hoarseness despite major safety concerns (laryngitis). It's also used as a "blood purifier" and to induce vomiting.

Figure 74: Queen's Delight.

Source: https://www.healthbenefitstimes.com/queens-delight/

Uses

- Digestive disorders.
- "Blood purification."
- Liver disease.
- Gallbladder disease.
- Constipation.
- Causing vomiting.
- Laryngitis.
- Hemorrhoids, when applied directly.
- Skin diseases, when applied directly.
- Other conditions.

Side Effects

When taken by mouth or applied to the skin, Queen's joy could be harmful. It contains compounds that have the potential to cause cancer. It may also trigger viruses that are already present in the body.

Queen's pleasure is highly irritating and can induce swelling anywhere on the body, including the skin, mouth, throat, and digestive tract. It can also cause nausea, vomiting, and diarrhea. Queen's delight can induce tongue and throat burning, painful urination, aches and pains, itching, rash, cough, sadness, weariness, and sweating in big doses.

Extracts

There has been no pharmacologic research on entire plant extracts.

Prostratin, a chemical component, is being studied for chemotherapy effects.

Dosage

The right dose of queen's delight is determined by a variety of criteria, including the user's age, health, and other circumstances. There is currently insufficient scientific data to define an appropriate dosing range for queen's joy. Keep in mind that natural products aren't always safe, and that dosages are crucial. Before using, be sure to read the product label and consult your pharmacist, physician, or other healthcare experts (Mati & de Boer, 2011).

Quince

Synonyms

Bedana, Cognassier, Coing, Coudonnier, Cydonia oblongata, Cydonia vulgaris, Marmelo, Membrillo, Pommier de Cydon, Pyrus cydonia, Quitte, Quittenbaum.

The quince (Cydonia oblonga) is the only member of the Cydonia genus in the Rosaceae family (which also contains apples and pears, among other fruits). It's a tree that produces a deciduous pome fruit that looks like a pear and matures to vivid golden-yellow color. Raw fruit is a food source. Landscape architects have grown the tree for its lovely pale pink blooms and other decorative attributes.

Uses

The seed has medicinal properties.

Quince is used to treating digestive problems such as stomach and intestine pain (gastrointestinal inflammation) and diarrhea as a powder, extract, or tea. Quince can also be used to treat coughs.

For injuries, swollen and painful joints, nipple soreness, and gashed or deeply wounded fingers, some individuals use quince directly as a compress or poultice. The eyes are soothed with a lotion.

Jam, jelly, marmalade, and pudding are all made with quince fruit. It's also utilized in the production of juice and wine.

- Digestive disorders.
- Diarrhea.
- Coughs.
- Stomach and intestinal swelling (inflammation).
- Skin injuries, when applied to the skin.
- Swollen and painful joints, when applied to the skin.
- Eye discomfort, when applied as a lotion.
- Other conditions.

Side Effects

There isn't enough information to know if quince is safe for medicinal use. The seeds contain cyanide, which suggests that quince seeds might not be safe.

Dosage

The right dose of quince is determined by several factors, including the user's age, health, and other circumstances. There is currently insufficient scientific evidence to define a suitable dosing range for quince. Keep in mind that natural products aren't always safe and that dosages are crucial. Before using, be sure to read the product label and consult your pharmacist, physician, or other healthcare expert.

Extracts

Quince is used to treat digestive problems such as stomach and intestine pain (gastrointestinal inflammation) and diarrhea as a powder, extract, or tea. Quince can also be used to treat coughs.

R

Ricinus Communis

Synonyms

African Coffee Tree, Arandi, Bi Ma Zi, Bofareira, Castorbean, Castor Bean, Castor Bean Plant, Castor Oil, Castor Oil Plant, Castor Seed, Erand, Eranda, Gandharva Hasta, Graine de Ricin, Huile de Ricin, Huile de Ricin Végétale, Mexico Weed, Palma Christi, Ricin, *Ricin Commun, Ricin Sanguin*, Ricinc, Ricino, *Ricinus communis, Ricinus sanguines*, Tangantangan Oil Plant, Wonder Tree (Li et al., 2015).

The castor bean or castor oil plant, Ricinus communis, is a perennial flowering plant in the Euphorbiaceae spurge family. It is the only species in the monotypic Ricinus genus and Ricininae subtribe. Modern genomic methods are currently being used to study the evolution of castor and its relationships with other species. It reproduces using a mixed pollination system that favors geitonogamy but can also be an out-crosser via anemophily (wind pollination) or entomophily (insect pollination) (insect pollination)

Figure 75: Leaves and flowers (male flowers on top).

Source: https://en.wikipedia.org/wiki/Ricinus

Its seed is the castor bean, which is not a bean despite its name (that is, the seed of many Fabaceae). Castor is native to the southeastern Mediterranean Basin, Eastern Africa, and India, but it's found all over the world (and widely grown elsewhere as an ornamental plant).

Castor seed yields castor oil, which has a wide range of applications. Between 40% and 60% of the oil in the seeds is high in triglycerides, primarily ricinolein. Racine, a water-soluble toxin found in lower concentrations throughout the plant, is also present in the seed.

Uses

- Before a colonoscopy, prepare your bowels. According to some studies, taking a single dosage of castor oil for bowel preparation before a colonoscopy is beneficial. Other bowel preparations, such as sodium phosphate or bisacodyl plus magnesium citrate, may be more helpful than castor oil.
- Constipation. When taken by mouth, castor oil acts as a stimulant laxative, which helps to relieve constipation.

- Contraception. A single dose of castor seeds without the outer coat (hulled) has been shown to operate as a contraceptive for up to 8-12 months (López-Malo et al., 2020).
- Eyes that are dry. According to several studies, utilizing eye drops containing castor oil may help persons with dry eyes.
- Stimulating full-term labor in women who are pregnant. In at least half of women at term who try it, a single 60 mL dosage of castor oil appears to trigger labor within 24 hours. There is also some evidence that women whose "water has broken" during a term pregnancy are more likely to go into labor and require a Cesarean section if they use castor oil.
- Syphilis.
- Arthritis.
- Skin disorders.
- Boils.
- Blisters.
- Swelling (inflammation) of the middle ear.
- Migraines.
- Softening cysts.
- Adhesive bowel obstruction.
- Warts.
- Bunions and corns.
- Promoting the flow of breast milk.
- Other conditions.

Side Effects

When consumed as a single dose, castor oil is safe for most people. Castor oil can induce stomach pain, cramping, nausea, and dizziness in some people.

When taken by mouth as a single dose, castor oil seeds with the outer coat removed (hulled) are safe. Castor oil eye drops are also safe when used for up to 30 days on the eye.

When consumed by mouth for a long time or in excessive dosages, castor oil is safe. When used for more than a week or in quantities of more than 15-60 mL per day, it may cause fluid and potassium loss in the body.

Taking the whole seed by mouth is quite dangerous. The castor seed's outer layer (hull) contains a lethal poison. Nausea, vomiting, diarrhea, stomach discomfort, dehydration, shock, blood cell destruction, severe fluid and chemical imbalances, liver, kidney, and pancreas damage, and death can all be caused by this outer layer. An adult can be killed by chewing just 1-6 entire seeds. Poisoning is less likely if the seed is ingested whole; nonetheless, quick medical assistance is still required.

Dosage

- Castor oil (15 mL) is widely used for constipation.
- To clean the gut before surgery or to examine the colon (colonoscopy), follow these steps: Adults and children over the age of 12 should take 15-60 mL of castor oil 16 hours before the surgery. 5-15 mL is usually used for children aged 2-11 years. 1-5 mL is widely used in children under the age of two.
- A variety of dose regimes have been utilized to begin delivery. Single doses of castor oil range from 5 to 120 mL. A 60 mL dose of fruit juice is typically used as a one-time dose. 5 mL in peppermint tea every 2 hours, 15 mL three times daily, 30 mL every 2 hours, 30 mL every 6 hours, 30 mL every 3 hours for three doses, 60 mL daily, and 60 mL daily for two days are some of the other dosing patterns that have been employed (Kujawska et al., 2017).

Raspberry Ketone

Synonym

The predominant fragrance compound of red raspberries is raspberry ketone, which is a natural phenolic substance. Raspberry ketone can be found in raspberries, cranberries, and blackberries, among other fruits. From coumaroyl-CoA, it is biosynthesized. It can be extracted from raspberries, giving about 1–4 mg per kilogram.

Because raspberry ketone is not abundant in nature, it is produced industrially using many processes using chemical intermediates. A crossing aldol condensation followed by catalytic hydrogenation is one technique to accomplish this. To make an,-unsaturated ketone, acetone is first condensed with 4-hydroxybenzaldehyde. The alkene is then converted to an alkane. Raspberry ketone is produced with a 99 percent yield using this two-step technique. Nickel boride is a less expensive hydrogenation catalyst that also has good selectivity for hydrogenation of the enone double bond.

Uses and Safety

To provide a fruity odor, raspberry ketone is sometimes used in fragrance, cosmetics, and as a food ingredient. It's one of the most expensive natural flavoring ingredients on the market. The natural substance can be expensive, costing up to $20,000 per kilogram. Synthetic raspberry ketone is less expensive, with prices ranging from a few dollars per pound to a quarter of the natural product's price.

The long-term safety of raspberry ketone supplementation is unknown, especially because human research has been limited. There are some concerns regarding its safety because it is chemically related to the stimulant synephrine. Cardiotoxic effects, as well as impacts on reproduction and development, are predicted by toxicological models. In addition, many nutritional supplements containing raspberry ketones contain additional chemicals, such as caffeine, which may have harmful consequences.

Raspberry ketone was categorized as generally recognized as safe (GRAS) by the US Food and Drug Administration in 1965 for use in tiny amounts to flavor foods.

Figure 76: Raspberry Ketone.

Source: https://www.whiteone.com/en/product/raspberry-ketone-supplements

Raspberry Leaf

Synonyms

The red raspberry is a shrub that produces a juicy, sweet berry that is extensively consumed. For millennia, red raspberry fruit and leaves have been utilized as medicine.

Some individuals take red raspberry leaf by mouth to help with labor and delivery, gastrointestinal (GI) problems including diarrhea, respiratory infections like the flu, and heart difficulties.

For a sore throat, a red raspberry leaf is gargled and rubbed to the skin for rashes.

Red raspberry fruit is consumed as well as processed into jams and other delicacies. In Europe, red raspberry leaf is used as a natural flavor in little amounts.

Uses

- The discomfort of childbirth. Taking red raspberry leaf did not shorten labor or minimize the need for pain relievers during birth.
- Abnormally heavy bleeding during menstrual periods (menorrhagia).
- Aging skin.
- Diabetes.
- Diarrhea.
- Flu (influenza).
- Gastrointestinal disorders.
- Heart disease.
- Heart failure and fluid build-up in the body (congestive heart failure or CHF).
- High blood pressure.
- Infection of the airways.
- Menstrual cramps (dysmenorrhea).
- Morning sickness.
- Preventing miscarriage.
- Skin rash.

- Sore throat (pharyngitis).
- Vitamin deficiencies.
- Other conditions.

Side Effects

When eaten in food amounts, red raspberry fruit is generally safe for most individuals. When consumed in larger doses as medicine, it is safe. There have been no known adverse effects from eating red raspberry fruit. However, a comprehensive assessment of red raspberry's safety has yet to be completed.

Dosage

The right dose of red raspberry depends on a variety of factors, including the user's age, health, and other circumstances. There is currently insufficient scientific evidence to define an optimal dosing range for red raspberry. Keep in mind that natural products aren't always safe and that dosages are crucial. Before using, be sure to read the product label and consult your pharmacist, physician, or other healthcare experts (Kunwar et al., 2016).

Red Clover

Synonym

The red clover is a type of plant. Medicine is made from flowers.

Red clover is used to treat menopause symptoms, weak and brittle bones, high cholesterol, and a variety of other ailments, however, there is no scientific proof to back up these claims.

Red clover is used as a flavor ingredient in meals and beverages.

Trifolium pratense, the red clover, is a herbaceous species of flowering plant in the bean family Fabaceae, native to Europe, Western Asia, and northwest Africa, but planted and naturalized in many other regions.

Uses.

- Bones that are weak and brittle (osteoporosis). The majority of studies demonstrate that eating red clover regularly does not improve bone density in women.

- Male-pattern baldness is a type of baldness that affects men (androgenic alopecia). Early study suggests that using a combination product containing red clover flower extract can help persons with hair loss grow more hair.
- Prostate enlargement (benign prostatic hyperplasia or BPH). According to a preliminary study, red clover supplements may help with some symptoms of an enlarged prostate. It appears to lessen urinating during the night.
- High cholesterol or other fats (lipids) (hyperlipidemia). In women with moderately raised cholesterol levels, the most study demonstrates that taking red clover extracts by mouth for 3 months to a year did neither reduce low-density lipoprotein (LDL or "bad") cholesterol or increase high-density lipoprotein (HDL or "good") cholesterol.
- Breast discomfort (mastalgia). Red clover may help alleviate cyclic breast discomfort and tenderness, according to preliminary research.
- Menopause symptoms. The majority of studies demonstrate that ingesting red clover by mouth for up to a year does not affect symptoms like hot flashes. However, some study suggests that taking a red clover supplement (Promensil, Novogen) may lessen the frequency of hot flashes. According to some studies, ingesting red clover can help with vaginal dryness (Kovach et al., 2009). Red clover has also been shown to help with psychological symptoms of menopause, such as depression and anxiety, according to other studies. However, not all research supports this.
- Asthma.
- Eczema is a skin condition that affects millions of people (atopic dermatitis).
- Breast cancer is a disease that affects women.
- Swelling (inflammation) of the lung's major airways (bronchitis).
- Burns.
- Cough.
- Irritability (dyspepsia).
- Uterine cancer is a type of cancer that affects the uterus's lining (endometrial cancer).
- Premenstrual syndrome (PMS) is a condition in which a (PMS).

- STIs (sexually transmitted infections) (STDs).
- Itchy, scaly skin (psoriasis).
- Cough caused by whooping cough (pertussis).
- Healing of wounds
- Other circumstances.

Dosage and Extracts

The right dose of red clover is determined by a variety of factors, including the user's age, health, and other circumstances. There is currently insufficient scientific evidence to define an optimal dosing range for red clover. Keep in mind that natural products aren't always safe and that dosages are crucial. Before using, be sure to read the product label and consult your pharmacist, physician, or other healthcare experts.

Clinical investigations have typically used extracts that have been standardized for isoflavone content. Biochanin A, formononetin, genistein, and daidzein are all present in these tablets. 7 A typical dose of total isoflavones is 40 to 80 mg per day (1 to 2 tablets). Higher doses (160 mg/day) have been employed in clinical trials, but their long-term safety has yet to be determined. 21 240 mg of crude plant extract equals 40 mg of isoflavones. There are several commercially available preparations.

Side Effects

When consumed at the levels found in food, red clover is generally safe for most individuals. When used at medical doses, it is completely safe. In some women, red clover can induce rashes, muscle pain, headaches, nausea, and vaginal bleeding (spotting).

When applied to the skin, red clover is considered safe when used in therapeutic doses.

Rhodiola

Synonym

Arctic Root, Extrait de Rhodiole, Golden Root, Hongjingtian, Hong Jing Tian, King's Crown, Lignum Rhodium, Orpin Rose, Racine d'Or, Racine Dorée, Racine de Rhadiola, Rhodiola rosea, Rhodiole, Rhodiole Rougeâtre, Rodia Riza, Rose Root, Rose Root Extract, Rosenroot, Roseroot, Rosewort,

Sedum rhodiola, *Sedum rosea*, Siberian Golden Root, Siberian Rhodiola Rosea, Snowdown Rose.

Rhodiola rosea is a plant that grows in cold climates throughout Europe and Asia. The root has been used in traditional medicine for a long time.

Rhodiola is a type of adaptogen. Adaptogens are natural chemicals that are thought to boost the body's tolerance to physical, environmental, and emotional stress. Rhodiola extract may aid in cell protection and heartbeat regulation.

Figure 77: Rhodiola has long been known as an adaptogen, a natural substance that increases your body's resistance to stress in non-specific ways.

Source: https://www.healthline.com/nutrition/rhodiola-rosea

Rhodiola is used to treat fatigue, anxiety, depression, stress, and a variety of other ailments, however, there is no scientific proof to back up any of these claims.

Side Effects

When taken by mouth: Rhodiola is possibly safe when taken for 6-12 weeks. Rhodiola might cause dizziness, dry mouth, or excessive saliva. There isn't enough reliable information to know if Rhodiola is safe to use long-term.

Dosage

Rhodiola has most often been used by adults in doses of 100-600 mg by mouth daily for up to 12 weeks. Speak with a healthcare provider to find out what dose might be best for a specific condition.

Extracts

Rhodiola rosea extracts have been used as a dietary supplement in healthy populations, including athletes, to help battle fatigue and depression by improving the body's inherent resistance to both physical and behavioral stresses. *Rhodiola rosea* extracts have recently been shown to exhibit anti-aging, anti-inflammation, immunostimulant, DNA repair, and anti-cancer properties in a variety of animal systems. The activity of *Rhodiola rosea* extracts has been examined mostly in conjunction with one of its bioactive components, salidroside. Both *Rhodiola rosea* extracts and salidroside have anti-cancer and anti-physiological activities molecular pathways.

Rooibos Red Bush Tea

Synonyms and Overview

The herbal infusion Rooibos tea is caffeine-free. It's made from Aspalathus linearis, a South African shrub.

The beverage has long been used as a natural treatment for a variety of ailments. Researchers are now looking into possible medical applications. Although nothing is conclusive at this time, several research have shown potential.

Uses

Tea can only be classified as such if it comes from the Camellia sinensis plant. While rooibos does not contain caffeine, it is brewed in the same way as tea and is frequently called to as such.

Traditional uses of rooibos included relieving colic in newborns. Wound healing, allergies, asthma, and sore throats were all treated with it (Karmakar et al., 2010).

Some of these applications are beginning to garner scientific support. Researchers have looked examined a variety of additional potential applications as well. However, most research has thus far been conducted in labs or on animals. To confirm what rooibos may be useful for, large human trials are required.

It has been used as an

- Antioxidant
- Anti-inflammatory
- Immunomodulator (affecting the functioning of the immune system)
- Liver protectant
- Brain protectant
- Antispasmodic (muscle spasms reliever)

Some of the best-researched uses of rooibos are as a treatment for:

- Heart disease
- Diabetes and metabolic syndrome
- Cancer

Side Effects

Most authorities agree that rooibos is "likely safe" for most people when consumed daily. There isn't enough evidence to say whether rooibos is safe for pregnant or lactating women.

One disadvantage of medicinal rooibos is that it may not be properly absorbed via the intestines. In addition, your liver may remove a large portion of it from your bloodstream before your body has a chance to utilize it.

This could imply that it will require a lot of money to make a difference. Large doses can cause liver damage in rare circumstances.

Dosage and Extracts

Infusions of Rooibos are made. Eight ounces of water are usually poured over one to two teaspoons of tea leaves.

According to some research, the chemicals in rooibos must steep for 10 minutes to reach medical levels.

Rose Hip

Synonyms

The rose hip, also known as rose haw or rose hep, is the rose plant's accessory fruit. It's usually red to orange, although some species go from dark purple to black. Rose hips form following flower pollination in the spring or early summer and ripen in the late summer and early fall (Kayani et al., 2015).

Apothecary Rose, Cherokee Rose, Cherokee Rose Musquée, Chinese Rosehip, Cynorhodon, Cynorhodons, Cynosbatos, Damask Rose, Dog Rose, Dog Rose Hips, Églantier, Fructus Rosae Laevigatae, Fruit de l'Églantier, Gulab, Heps, Hip, Hip Fruit, Hip Sweet, Hipberry, Hop Fruit, Jin Yin Zi, Jinyingzi.

Figure 78: Rose hips from Rosa rugosa (beach rose).

Source: https://en.wikipedia.org/wiki/Rose_hip#/media/File:Rose_hips.jpg

Uses

- Osteoarthritis. Rose hip extract, taken by mouth alone or in combination with other natural medicines, can help people with osteoarthritis reduce pain and stiffness while also improving function.

- Following surgery, there is pain. Taking a single dosage of the rose hip extract by mouth before a C-section can assist to minimize pain and the need for pain medication following the procedure.

Rose hip is being considered for a variety of additional uses, but there isn't enough trustworthy information to say whether it will be beneficial.

Side Effects

When taken by mouth, a rose hip extract is probably safe when consumed in food proportions. When taken in bigger doses as medicine, rose hip from Rosa canina is believed to be safe. When consumed in bigger doses as medicine, rose hip from Rosa damascena is likely safe. There isn't enough solid data to say whether rose hip from other types of roses is safe in medicinal doses. Rose hip can cause diarrhea and other negative effects.

When applied to the skin, there isn't enough solid data to tell if the rose hip is safe or what the potential adverse effects are.

Extracts

Rosehip extract contains polyphenols and anthocyanins, which are believed to ease joint inflammation and prevent joint damage. It's also rich in vitamin C, which has antioxidant properties. Antioxidants are substances that can override harmful molecules (free radicals) which are produced within your cells and which may cause tissue damage or disease. Other studies have found that it can reduce the production of specific enzymes that break down cartilage (Athar & Nasir, 2005).

Dosage

Rose hip has most often been used by adults in doses of up to 5 grams by mouth daily for up to 12 weeks. Speak with a healthcare provider to find out what dose might be best for a specific condition

Rosemary

Synonyms

Salvia rosmarinus, also known as rosemary, is a Mediterranean shrub with aromatic, evergreen, needle-like leaves and white, pink, purple, or blue flowers. It was previously known under the scientific name Rosmarinus officinalis, which is now a synonym.

It belongs to the Lamiaceae family, which also comprises many other medicinal and culinary plants. The word "rosemary" comes from the Latin ros marinus, which means "salt of the sea" ("dew of the sea"). The plant is also known as anthos, which comes from the ancient Greek word o, which means "flower." The root structure of rosemary is fibrous.

Uses

Rosemary may have a variety of health benefits.

Compounds that are anti-oxidants and anti-inflammatory

Rosemary is high in antioxidants and anti-inflammatory chemicals, which are known to promote blood circulation and stimulate the immune system.

According to laboratory research, rosemary is high in antioxidants, which help to neutralize damaging particles known as free radicals.

Enhancing digestion

Rosemary is commonly used to relieve dyspepsia in Europe. In reality, rosemary has been certified by Germany's Commission E for the treatment of dyspepsia. It should be noted, however, that there is currently no credible scientific data to back up this assertion.

Memory and focus are improved.

The perfume of rosemary can increase a person's concentration, performance, speed, and accuracy, as well as mood, according to research published in Therapeutic Advances in PsychopharmacologyTrusted Source.

Protection for the brain

Rosemary has been discovered to be beneficial to the brain by scientists. Rosemary includes a compound called carnosic acid, which can protect the brain from free radical damage.

Side Effects

Rosemary is generally safe in small doses. Extremely excessive doses, on the other hand, can cause major side effects, albeit this is uncommon.

coma pulmonary edema vomiting spasms (fluid in the lungs)

Due to the risk of miscarriage from high dosages of rosemary, it is not recommended for pregnant women to consume any supplemental rosemary (Kassam et al., 2010).

Dosage

Rosemary is safe when eaten by mouth at the levels found in foods. When taken by mouth as a medicine in doses up to 6 grams per day, rosemary leaf is generally safe for most people. Taking undiluted rosemary oil or excessive volumes of rosemary leaf, on the other hand, is dangerous. Large doses of rosemary can induce nausea, uterine bleeding, renal irritation, increased light sensitivity, skin redness, and allergic responses in certain people.

When applied to the skin for medical purposes, rosemary oil is generally harmless for most people. Some people may get allergic reactions as a result of it.

When inhaled as aromatherapy for medicinal purposes, Rosemary is safe for most people.

Extracts

The use of extracts from rosemary (*Rosmarinus officinalis*) as food preservatives is well established. A broad range of beneficial health effects can be attributed to rosemary, such as antidepressant, antihypertensive, antiproliferative, antibacterial, antiatherogenic, hypocholesterolemic, hepatoprotective, and anti-obesity properties.

Rosemary extracts derived from *Rosmarinus officinalis L.* contain several compounds which have been shown to possess antimicrobial, antioxidative, anti-inflammatory, antiviral, and anti-tumor functions

Royal Jelly

Synonyms

Bee Saliva, Bee Spit, Gelée Royale, Honey Bee Milk, Honey Bee's Milk, Jalea Real, Lait des Abeilles, Royal Bee Jelly.

Worker honeybees make royal jelly, which is a milky substance. It usually comprises 60 to 70% water, 12 to 15 percent proteins, 10 to 16 percent sugar, 3 to 6 percent lipids, and 2 to 3 percent vitamins, salts, and amino acids. Its make-up changes depending on location and climate (Jiang et al., 2019).

The name of this product comes from the fact that bees use it to raise queen bees. Royal jelly is used as a medication by some people. Royal jelly is not to be confused with bee pollen, beeswax, bee venom, or propolis.

Figure 79: Developing queen larvae surrounded by royal jelly.

Source: https://en.wikipedia.org/wiki/Royal_jelly

Uses

Worker bees generate royal jelly, a milky-white fluid that aids in the growth of the queen bee. Apis mellifera, Bee Saliva, Bee Spit, Gelée Royale, Honey Bee Milk, Honey Bee's Milk, Jalea Real, Lait des Abeilles, or Royal Bee Jelly are all names for royal jelly.

Apitherapy, bee pollen, and bee venom are not to be confused with royal jelly.

In alternative medicine, royal jelly has been touted as a possible treatment for menopausal symptoms. To treat this ailment, royal jelly may have been blended with other plants or extracts in a specialized recipe.

Hay fever has also been treated with royal jelly. Royal jelly, on the other hand, has been demonstrated in studies to be ineffective in the treatment of this illness.

Premenstrual syndrome, infertility, excessive cholesterol, diabetic foot ulcers, stomach ulcers, pancreatitis, liver or renal illness, asthma, skin diseases, and a variety of other conditions are among the unproven uses.

It is unknown whether royal jelly can be used to treat any medical ailment. The FDA has not authorized this product for medical use. Royal jelly should not be used in place of any prescription that your doctor has prescribed for you (Jima & Megersa, 2018).

Royal jelly is often sold as an herbal supplement. There are no regulated manufacturing standards in place for many herbal compounds and some marketed supplements are contaminated with toxic metals or other drugs. Herbal/health supplements should be purchased from a reliable source to minimize the risk of contamination.

Doses

BY MOUTH: Menopause Symptoms: For the past eight weeks, 1000 mg of royal jelly has been taken daily. For up to 12 weeks, one or two capsules of a product combining royal jelly and flower pollen were taken. Daily for four weeks, two capsules of a product containing royal jelly, evening primrose oil, damiana, and ginseng were consumed.

Side Effects

When eaten by mouth, royal jelly is generally considered safe for most people when taken in the recommended doses. Doses of up to 4.8 grams per day have been taken safely for up to a year. Royal jelly can trigger major allergic responses in persons who have asthma or allergies.

When used on the skin, royal jelly is completely harmless when used correctly. When applied to the scalp, however, it has caused inflammation and an allergic response.

Respiratory Syncytial Virus (RSV)

Respiratory syncytial virus (RSV), also known as human respiratory syncytial virus (hRSV) and human orthopneumovirus, is a highly contagious virus that causes respiratory tract infections. It's a single-stranded, negative-sense RNA virus that gets its name from the huge cells called syncytia that form when infected cells fuse together.

RSV is the most common cause of respiratory hospitalization in babies, and reinfection is common later in life: it is a serious disease that affects people of all ages. During the cold winter months, infection rates are often greater, resulting in bronchiolitis in newborns, common colds in adults, and more serious respiratory infections such as pneumonia in the elderly and immunocompromised

RSV is spread by infected air droplets and can produce epidemics in the community as well as in hospitals. The virus infects the epithelial cells of the upper and lower airways after an initial infection through the eyes or nose, causing inflammation, cell destruction, and airway blockage. RSV can be detected and diagnosed using a variety of approaches, including antigen testing, molecular testing, and viral culture. While hand washing and avoiding close contact with infected people are the most effective ways to prevent RSV infection in high-risk infants, a preventative drug called palivizumab is also available. There is currently no vaccination available for RSV although many are under development.

Symptoms

Signs and symptoms of respiratory syncytial virus infection most commonly appear about four to six days after exposure to the virus. In adults and older children, RSV usually causes mild cold-like signs and symptoms. These may include:

- Congested or runny nose
- Dry cough
- Low-grade fever
- Sore throat
- Sneezing
- Headache

Causes

The respiratory syncytial virus enters the body via the eyes, nose, or mouth. It quickly spreads through the air on infected respiratory droplets. If someone with RSV coughs or sneezes near you, you or your kid may become infected. The virus can also be transmitted by direct touches, such as shaking hands.

The virus can survive for hours on hard surfaces including worktops, crib rails, and toys. Handling your mouth, nose, or eyes after touching a contaminated object increases your chances of contracting the virus.

During the first week or so following infection, an infected person is most contagious. However, in infants and those with low immune systems, the virus may continue to spread even after symptoms have subsided for up to four weeks.

Prevention

There is no vaccine available for the respiratory syncytial virus. However, the following lifestyle behaviors can aid in the prevention of the spread of this infection:

- Hands should be washed often. Teach your kids the value of hand washing.
- Avoid becoming exposed. When you cough or sneeze, cover your mouth and nose. Avoid exposing your kid to persons who have fevers or colds (Helmstädter & Staiger, 2014).
- Maintain a clean environment. Make sure the kitchen and bathroom countertops, as well as the doorknobs and handles, are clean. Throw away used tissues as soon as possible.
- Drinking cups should not be shared with others. When you or someone else is unwell, use your glass or disposable cups. Each person's cup should be labeled.
- Don't light up. Babies who are exposed to cigarette smoke are more likely to contract RSV and experience more severe symptoms. If you must smoke, never do it in your home or car.
- Toys should be washed regularly. This is especially important when your child or a playmate is ill.

Reverse Transcriptase

A reverse transcriptase (RT) is an enzyme that produces complementary DNA (cDNA) from an RNA template, through a process known as reverse transcription. Reverse transcriptases are employed by viruses to reproduce their genomes, retrotransposon mobile genetic elements to proliferate inside the host genome, and eukaryotic cells to lengthen the telomeres at the ends of their linear chromosomes. Contrary to popular assumption, the process does not break the fluxes of genetic information specified by the conventional core dogma, because information transfers from RNA to DNA are specifically allowed.

Retroviral RT has three biochemical activities that occur in sequential order: RNA-dependent DNA polymerase activity, ribonuclease H (RNase H), and DNA-dependent DNA polymerase activity. These actions, when combined, allow the enzyme to convert single-stranded RNA into double-stranded cDNA. This cDNA can then integrate into the host genome in retroviruses and retrotransposons, from which new RNA copies can be produced by host-cell transcription. In the laboratory, the same sequence of processes is commonly employed to convert RNA to DNA for use in molecular cloning, RNA sequencing, polymerase chain reaction (PCR), or genome research (Agbogidi & Azagbaekwe, 2013).

The enzymes are expressed and employed by viruses that use reverse transcription as part of the replication process. Retroviruses, for example, employ the enzyme to reverse-transcribe their RNA genomes into DNA, which is then integrated into the host genome and duplicated alongside it. Reverse-transcribing DNA viruses, such as hepadnaviruses, can employ RNA as a template in the assembly and formation of DNA strands. This enzyme is used by HIV to infect people. The viral genome would not be able to incorporate into the host cell without reverse transcriptase, resulting in replication failure.

S

Sandalwood

Sandalwood is an evergreen tree native to India and Indonesia that reaches a height of 8 to 12 meters and a diameter of 2.5 meters. The little blooms have numerous short stalks, and the bark is smooth and gray-brown in hue. White saunders oil, santal oil, yellow or white sandalwood oil, and East Indian sandalwood oil are all names for sandalwood. Sandalwood oil has a sweet, woody scent and is often used in aromatherapy, perfumes, cosmetics, and soaps as a fragrance. It's also utilized as a flavoring in meals and drinks. Because of its solid nature, the wood has long been prized for carving.

In alternative medicine, sandalwood and its essential oil are highly appreciated. It's been used for centuries to treat a range of ailments. Although most of its traditional uses have not been scientifically validated, some study suggests that sandalwood may be useful in the treatment of certain health issues.

Figure 80: Sandalwood essential oil works wonders as an antiseptic. It is suitable for both internal and external use. When taken orally, it aids in the prevention of infection in internal wounds and ulcers.

Source: https://www.lybrate.com/topic/benefits-of-sandalwood-oil-and-its-side-effects

Sandalwood oil has long been used in traditional medicine as an antibacterial and astringent, as well as for the management of headaches, stomachaches, and urinary and vaginal issues. Sandalwood essential oil, emulsion, or paste is used to cure inflamed and eruptive skin problems in India. The oil has been used as a diuretic and moderate stimulant in ancient Ayurvedic medicine, as well as for skin softening (Bussmann & Sharon, 2006). The leaves and bark were utilized to alleviate dermatitis, lice, skin irritation, and sexually transmitted infections by the early Hawaiians. Sandalwood oil has also been shown to resist Tetranychus urticae, an agricultural pest (two-spotted spider mite).

A dose of 1 to 1.5 g daily for no more than 6 weeks is indicated for the management of urological disorders. Sandalwood oil should be dosed in a coat that is resistant to gastrointestinal fluids. Ingestion of the oil in its natural state is not recommended. Although sandalwood oil is generally thought to be non-irritating to human skin, it can induce skin inflammation. The Flavor and Extract Manufacturers' Association has designated sandalwood oil as a generally recognized safe (GRAS) flavoring additive in food, and the US Food and Drug Administration (FDA) has designated it as a natural flavoring.

Saccharum Officinarum

This is commonly called sugarcane. Sugarcane is a perennial grass that grows to be quite huge. It develops thick culms that can reach a height of 2 to 3 meters, but with longer growing seasons, the plant can reach even higher heights. It is farmed mostly for the sap in the stems, which is used to make sugar, alcohol for fuel, and other products in the tropics. The plant is also a valuable wax and a substantial source of biomass. It has grown since the dawn of time. Some kinds have a dense covering of siliceous hairs on their leaves and sheaths. Cane cutters favor kinds with few hairs or charred cane since these hairs penetrate their skin and are quite unpleasant.

A delicious sap is found in the stems. This sap is extracted by crushing them, and it is subsequently used in a variety of ways. It can be made into a sweet, refreshing drink, although it is most usually condensed into syrups or dried into sugar, molasses, and other products. Fresh stems' cores are nibbled for a pleasant delight.

Sore eyes are treated with leaf ash. Sore throats are treated with stem juice. Snakebite and poison arrow wounds are treated with the delicious nectar in the stem. It is used to cure curare toxicity when combined with an infusion of 'wallaba' (Eperua sp.). Urinary problems are treated with a decoction of the young leaves.

Sugarcane juice is popularly known in India as a nutritional drink and is regarded as a unique source of varied concentrations of various hydrophilic components, as well as substantial biological activity. The richest source of phenolic compounds, such as phenolic acids, flavonoids, and various glycosides, is sugarcane juice and its unprocessed derivatives, such as brown sugar, molasses, and jaggery.

Read the labels on products to see what they include if you want to try sugarcane juice. A small amount of sugarcane juice is blended with water and sugar in some goods, but there is no fiber. Unless a product is made entirely of unprocessed sugarcane juice, it contains added sugar, which has been linked to a slew of negative health consequences. Sugary foods and beverages are higher in calories, which contributes to obesity. Sugar of any kind allows bacteria to proliferate in the mouth, increasing the risk of cavities. Diabetes and fatty liver disease are two other possible side effects. Excess sugar consumption boosts triglyceride levels, which increases the risk of cardiovascular disease. Sugar has also been related to greater blood pressure and inflammation, both of which are associated to heart disease (Blancke, 2016).

Sapindus Rarak

Sapindus rarak is a deciduous tree with a height of 42 meters. The diameter of the straight bole can be up to 100cm. The tree's fruits, which are a great source of saponins that can be used as soap and insecticides, are picked from the wild. It is used as a pioneer species in Thailand's reforestation programs. The fruits are used to cure scabies on the outside.

The fruits are high in saponins and are used as a soap alternative by the locals. The drupes were once employed as varnish, floor wash, and soap because they lather when mixed with water. Buttons and beads are made from fruits and seeds. Insecticides are made from fruits. The sapwood is yellowish-white and the heartwood is yellowish-grey. The wood is tough but not long-lasting. Furniture, sawing boards, plywood, and boards are all made from it (Gupta et al., 2018).

Saraca Indica

Ashoka, also known as *Saraca indica*, is a rainforest tree native to India's Deccan Plateau and the Western Ghats. It blooms in clusters of orange and yellow flowers that turn crimson before falling. This tree is linked to the love of God Kama and is regarded as one of the most sacred trees in Indian beliefs. In Sanskrit, the word 'Ashoka' means 'no sorrow,' and various portions of this tree are frequently employed for therapeutic purposes.

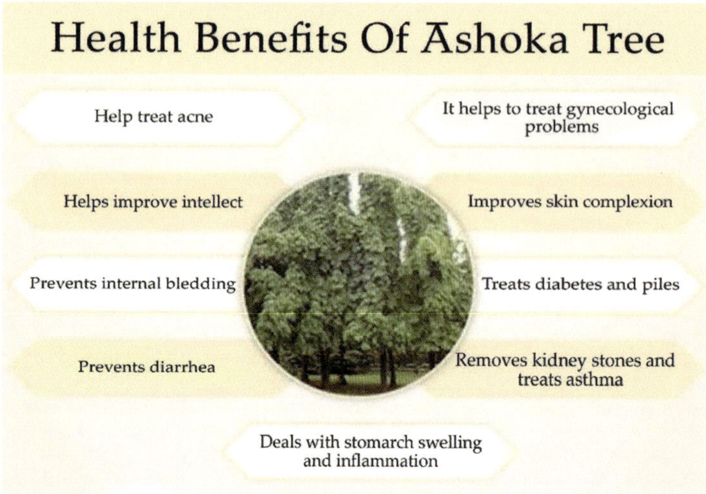

Figure 81: Ashoka, also known as Saraca indica, is a rain forest tree native to India's Deccan Plateau and Western Ghats. It blooms in clusters of orange and yellow flowers that turn crimson before falling.

Source: https://www.lybrate.com/topic/ashoka-tree-uses-benefits-and-side-effects

The Ashoka tree is revered in Ayurveda for its spiritual qualities as well as its numerous health advantages. The bark of the Ashoka tree is largely used to treat menstrual disorders in women, such as menstrual cramps, stomach pain, and uterine spasms. The skin of the Ashoka tree is high in tannins, glycosides, and flavonoids, all of which act as uterine tonics. The roots and seeds of the Ashoka tree are used to cure skin diseases such as acne, psoriasis, and eczema.

The Ashoka tree can provide a variety of critical nutrients. Tannins, glycosides, essential oils, and calcium are all present. It also comprises carbonic components of carbon and iron, as well as ketosteroid, which is found in the tree's bark. The intake of Ashoka tree herbs aids in the removal of toxins from our blood, providing good skin benefits. Herbs from the Ashoka tree detoxify the blood and help to eliminate allergy symptoms and other skin problems. Extracts from the Ashoka tree can also be used to heal burns and skin irritations.

To avoid internal bleeding, consume a few drops of water mixed with crushed Ashoka flower petals. Hemorrhagic dysentery is also prevented by eating ashoka flowers. As a result, Ashoka blossoms might assist you in maintaining your health. Diabetes patients can utilize the dried flowers of the Ashoka tree to help them manage their condition (Griffiths et al., 2002). The Ashoka flowers aid in the regulation of blood sugar levels. The Ashoka tree provides herbal compounds that aid in the treatment of piles. Internal piles can be treated with a decoction made from Ashoka tree bark. Analgesic characteristics of Ashoka tree extract aid in pain relief. If you have joint pain, a paste made from the bark can be applied to the affected areas to provide relief.

Dosages for syrup is 10 to 15 milliliters, for the flower powder is 1 to 3 grams, for the seed powder is 1 to 3 grams, and for the bark, a decoction is 15 to 30 ml. Despite the numerous health benefits of the Ashoka tree, pregnant women should avoid consuming its products because they may cause difficulties. Before consuming this tree's products, people who are allergic to particular foods should visit a doctor.

Stachytarpheta Jamaicensis

Stachytarpheta jamaicensis is a perennial herb that grows up to 1 meter tall and is well-branched. At the base, the stems grow woody. The herb is typically harvested from the wild for local use and has a wide range of medical benefits. In India, Southeast Asia, and the Caribbean, the plant is used as a food crop and medicinal herb. It is sometimes used as a decorative plant.

The twig tips are utilized as a seasoning and a pot herb. The green leaves can be cooked on their own or combined with other vegetables. The mixture is then served with a staple dish like rice or ugali, and coconut milk or pounded groundnuts are added. The leaves have previously been employed as a tea adulterant. The dried leaves are used to make tea on their own or to adulterate Chinese tea (Fitzgerald et al., 2020).

Emetic, expectorant, sudorific, and tonic properties are said to be present in the plant. Malaria, yellow fever, amenorrhea, syphilis, and gonorrhea have all been treated with it locally. Vomiting, diarrhea, diabetes, low blood pressure, and fractures are all treated using the whole plant. To treat grippe and headaches, as well as liver diseases, a decoction of the entire plant is used as a bath. To cure nervous problems, a tincture of the herb in rum is used as a lotion. The seedling has a purgative effect. The root has a depurative effect.

Purgative, hypotensive, astringent, and cholagogue, properties are found in the leaves. High blood pressure is treated with an infusion. The juice from the leaves is used to relieve colic and dysentery pain. The crushed leaves are used to make a stew that is intended to treat gonorrhea and reduce sweating. To treat cutaneous eruptions, a decoction containing sodium sulfate is consumed. Externally, the leaves are used as a resolutive cataplasm and to heal skin burns. To treat earache, the leaves' juice is poured into the ears.

If you are allergic to aspirin, do not take it. Also, do not take it if you are pregnant.

Symphytum Officinale

Also called Healing Herb, Herbe aux Charpentiers, Ass Ear, Black Root, Blackwort, Bruisewort, Common Comfrey, Consolidae Radix, and Confound, Comfrey is a type of herb. Despite the presence of deadly compounds known as pyrrolizidine alkaloids (PAs), the leaf, root, and root-like stem (rhizome) are used to create medication.

Comfrey is taken by mouth for cancer, bronchitis, cough, bloody urine, diarrhea, stomach ulcers, heavy menstrual cycles, and chest pain, despite safety concerns (angina). It can also be gargled to treat gum disease and a sore throat. Ulcers, rheumatoid arthritis, bruises, muscle discomfort, wounds, varicose veins, gout, and fractures are treated with comfrey applied to the skin. When applied to the skin, the compounds in comfrey may help to heal wounds and reduce inflammation. Comfrey, on the other hand, has poisonous compounds that can be absorbed through the skin (Eddouks et al., 2017).

Lower or upper back discomfort appears to be relieved by applying a comfrey extract ointment to the affected area for 5 days. Additionally, using a lotion comprising comfrey extract and methyl nicotinate to the afflicted area for 5 days seemed to reduce back pain while resting or moving. In persons with knee osteoarthritis, use a comfrey extract ointment to the affected area for 3 weeks or a special cream comprising comfrey extract, tannic acid, Aloe vera gel, eucalyptus oil, and frankincense oil to the affected area for 6-12 weeks helps to reduce discomfort.

Comfrey

Botanical name: *Symphytum officinale*

Family: Boraginaceae

Parts used: leaves, root

Taste: sweet, salty

Energetics: cooling, moistening

Plant Properties: cell proliferant, nutritive, demulcent, expectorant, vulnerary

Plant Uses: External use for healing of clean wounds, broken bones, pulled ligaments/sprains, varicose veins, burns, and hernias. Internal use with caution for coughs, ulcers, internal bleeding, hemorrhoids. Garden amendment.

Plant preparations: infusion, decoction, oil, poultice

Figure 82: Comfrey has been found to contain pyrrolizidine alkaloids (PAs), which are chemicals that have been linked to liver damage in humans. Over 600 distinct PAs have been discovered in over 6,000 different plants.

Source: https://www.herbalremediesadvice.org/benefits-of-comfrey.html

When consumed by mouth, comfrey is almost certainly unsafe for anyone. It includes compounds known as pyrrolizidine alkaloids (PAs) that can harm the liver, and lungs, and cause cancer. All oral comfrey products should be taken off the market, according to the FDA. When applied to the skin, comfrey is POSSIBLY SAFE for most people when administered in modest doses to unbroken skin for fewer than 10 days. It's crucial to keep in mind that comfrey's harmful compounds might pass through the skin. As a result, using comfrey to break skin or applying significant amounts to the skin for more than 6 weeks is POSSIBLY UNSAFE.

In scientific studies, the following doses have been investigated: For skin applications, for back discomfort, 4 grams of a specialized ointment containing 35% comfrey root extract was administered three times daily for five days. A 4-gram cream containing 35 percent comfrey and 1.2 percent methyl nicotinate was also applied three times each day for five days (Dhama et al., 2018). For osteoarthritis, 2 grams of a special ointment containing 35% comfrey root extract was administered three times daily for three weeks to the knee. In addition, for 6-12 weeks, roughly 3.5 grams of a cream containing comfrey extract, tannic acid, aloe vera gel, eucalyptus oil, and frankincense oil was applied three times daily to the knee. For sprains, about 2 grams of a particular ointment containing 35% comfrey extract was administered four times daily for eight days to ankle sprains.

Seasonal Affective Disorder

Seasonal affective disorder (SAD) is a type of depression that is linked to seasonal changes. SAD occurs at roughly the same time each year. If you're like the majority of SAD sufferers, your symptoms begin in the fall and last through the winter, sapping your energy and making you cranky. In the spring and early summer, SAD is less likely to produce depression. Seasonal affective disorder symptoms typically develop in late fall or early winter and disappear over the warmer months of spring and summer. People with the opposite pattern are more likely to develop symptoms in the spring or summer. In any scenario, symptoms may begin mildly and rise to more serious levels as the season advances.

The seasonal affective disorder has yet to be identified as a specific cause. The following are some of the factors that may come into play: Your biological clock is ticking (circadian rhythm): Winter-onset SAD may be caused by the decreasing amount of sunlight in the fall and winter. The lack of sunlight may cause your body's internal clock to malfunction, resulting in despair (Bhattacharyya, 2015). Serotonin levels: A decline in serotonin, a mood-altering brain chemical (neurotransmitter), may have a role in SAD. Reduced sunshine can lead to a drop in serotonin levels, which can lead to depression. Levels of melatonin: The changing of the seasons can throw off the body's melatonin balance, which affects sleep patterns and mood.

Women are diagnosed with seasonal affective disorder more frequently than males. SAD is more common in younger persons than it is in older adults. The seasonal affective disorder can be caused by several factors, including SAD sufferers are more likely to have blood relations who suffer from SAD or another form of depression. Having serious depression or bipolar disorder: If you have one of these disorders, your depression symptoms may intensify seasonally. Living far from the equator: People who dwell far north or south of the equator appear to be more susceptible to SAD. This could be owing to the shorter days in the winter and the longer days in the summer.

Adaptogenic herbs boost mood, increase mental and athletic productivity, and improve the body's response to stress. These herbs can be taken tonically to assist relaxation and balance mood and energy, or utilized in the short term to overcome tiredness and push through difficult conditions. For millennia, Ashwagandha (Withania somnifera), a nightshade family member, has been used in Ayurvedic and Unani medicine as a "rasayana" (rejuvenator) for a variety of ailments, including insomnia, debility, infertility, skin and digestive problems, and a variety of infections. It's moderate enough to take before night (the conventional technique), but not so sedating that it can't be used during the day to support a balanced stress response.

For millennia, Rhodiola (R. rosea), a crassulaceous plant endemic to cold, severe climes from Tibet to Siberia and Scandinavia, has been used to improve endurance, longevity, and a happy attitude. Rhodiola rhizome activates noradrenaline, serotonin, dopamine, and acetylcholine receptors in the brain, raises serotonin by inhibiting MAO enzymes, and increases endogenous endorphins, according to in vitro and animal research. A short human experiment found that rhodiola extract improved depression symptoms moderately.

S-adenosylmethionine Scu Subcutaneous

S-adenosylmethionine (commonly known as SAMe) is a synthetic version of a molecule found in the human body. SAMe is a potentially beneficial aid in lowering depression symptoms and treating osteoarthritis, according to alternative medicine. SAMe may be useful for treating liver disease in pregnant women, fibromyalgia symptoms, AIDS nerve difficulties, and sexual dysfunction in males with depression. Treatment of liver illness, heart disease, psychosis, anxiety, tendinitis, chronic back pain, severe headaches, convulsions, menstrual problems, and chronic fatigue syndrome are examples of non-researched usage (Devkota & Watanabe, 2020).

It is unknown if SAMe is effective in the treatment of any medical problem. The FDA has not authorized this product for medical use. SAMe should not be used in place of any prescription that your doctor has prescribed for you. SAMe is frequently marketed as a natural supplement. Many herbal compounds have no defined production requirements, and some advertised supplements are tainted with harmful metals or other medications. To reduce the possibility of contamination, herbal/health supplements should be acquired from a reputable source.

Steven Foster

Steven Foster, a herbalist, author, and photographer, has spent more than four decades photographing and researching herbs all over the world, from the Amazon rainforest to the highlands of Vietnam. Foster has approximately 900 photo-illustrated pieces in a variety of publications. Since 1998, he has been a member of the American Botanical Council's Board of Trustees. His photographs and articles appear often in HerbalGram, where he also works as a Contributing Editor. Steven has been involved with ABC since its inception in 1988.

Figure 83: Foster is the president of Steven Foster Group, Inc., a company that offers consultancy, editorial, and stock photography services.

Source: http://abc.herbalgram.org/site/PageServer?pagename=Steven_Foster

Foster has become one of the country's most respected specialists in the field, with over a dozen books to his credit, in the nearly thirty years since he began his study of herbs and plants. He is also a well-known medicinal plant photographer, with over 120,000 photographs in his stock library. From his debut book, Herbal Bounty: The Gentle Art of Herb Culture, published in 1984, to the several Peterson's Field Guides on medicinal and toxic plants that he has co-authored, Foster's knowledge of helpful herbs is on display in several of his books (Douwes et al., 2008).

Tyler's Honest Herbal: A Sensible Guide to the Use of Herbs and Related Remedies, a standard reference that Foster coauthored with Varro Tyler in its fourth edition, likewise emphasizes science. This book delves into the pharmaceutical science behind herbal treatments in a way that is accessible to the general public.

Serum Glutamic Pyruvic Transaminase

This enzyme is generally found in the liver and cardiac cells. When the liver or heart is injured, SGPT is released into the bloodstream. As a result, when the liver is damaged (for example, from viral hepatitis) or when the heart is injured, the blood SGPT levels rise (for example, from a heart attack). SGPT levels can also be raised by several medicines. Alanine aminotransferase is another name for SGPT.

ALT is a component of the AST/ALT ratio and is routinely tested clinically as part of liver function testing. It is generally always quantified in international units/liter (IU/L) or kat when used in diagnostics. While sources differ on individual patient reference range numbers, the usual reference range for experimental investigations is 0-40 IU/L.

When excessive ALT levels are detected in the blood, other enzymes can be measured to narrow down the likely causes. By monitoring alkaline phosphatase, for example, high ALT levels related to hepatocyte injury can be differentiated from bile duct issues. When the aspartate transaminase (AST) is greater than ALT, myopathy-related elevations in ALT should be considered; muscle enzymes, especially creatine kinase, can be used to further investigate the likelihood of muscle illness generating elevated liver values (Do et al., 2019).

Because significantly raised levels of ALT (SGPT) can indicate the presence of various medical issues such as viral hepatitis, diabetes, cardiomyopathy, liver damage, bile duct problems, autoimmune disease, or myopathy, ALT is frequently used as a screening tool for liver issues. Dietary choline shortage might also induce elevated ALT. However, high ALT levels do not always imply that there are medical issues. The fluctuation of ALT levels throughout the day is natural, and they can also rise in reaction to hard physical activity.

Sesquiterpene Lactones

Sesquiterpene lactones (SLs) are a type of sesquiterpene that has a lactone ring attached to it. They are most commonly seen in the Asteraceae family of plants (daisies, asters). Umbelliferae (celery, parsley, carrots) and Magnoliaceae are two other plant families possessing SLs (magnolias). SLs are a rich source of pharmaceuticals because they are a collection of colorless, lipophilic solids. In grazing livestock, they can be allergic and poisonous, causing serious neurological issues in horses. Some of them can also be found in corals like Maasella edwardsi.

Traditional or folk medicine makes extensive use of compound-rich plants, such as those in the Asteraceae family, many of which are commercially available as herbal medicines. These are notably high in a kind of natural terpenoids known as sesquiterpene lactones, which are sometimes regarded the active ingredients in medicinal plants. Sesquiterpene lactones are terpenes with a 15-carbon basic structure (thus the prefix sesqui-) that originate from the biosynthesis of three isoprene units with a cyclical structure and a fused -methylene—lactone ring.

Compounds from several of these subclasses have a diverse set of biological functions. These include antimalarial, hepatoprotective, antioxidant, antibacterial, neuroprotective, and immune-stimulant characteristics, and they vary from anticancer to anti-inflammatory. In terms of the structure/activity relationship of these compounds, it appears that the -methylene—lactone nucleus plays a key role in practically all of their known biological effects, including anti-inflammatory, anti-trypanosomal, and cytotoxic effects (Cummings & Olsen, 2011). Sesquiterpene lactone activity appears to be influenced by other structural moieties. Electrophilic locations linked with medium/high lipophilicity, for instance, appear to promote anti-mycobacterial action, whereas a double bond Exo to the cyclopentenone ring appears to favor anti-inflammatory function.

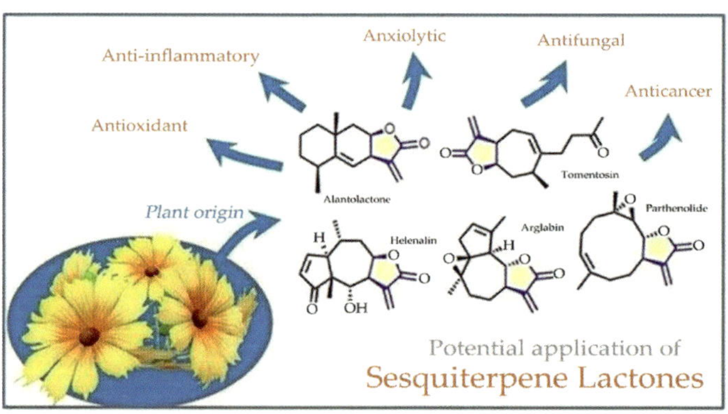

Figure 84: Sesquiterpene lactones, a diverse group of terpenoids identified from Asteraceae species, have a wide range of biological effects, and several of them, such as artemisinin, are now commercially available.

Source: https://www.mdpi.com/2076-3417/10/9/3001/htm

Systemic Lupus Erythematosus

Lupus, also known as systemic lupus erythematosus, is an autoimmune illness with numerous forms (SLE). It happens when your immune system attacks tissues and organs all over your body. Lupus can cause inflammation and damage to the kidneys, skin, joints, brain, blood vessels, lungs, heart, and other parts of the body. It is known to cause major health problems (including kidney disease, rheumatoid arthritis, and neurological dysfunction). While lupus symptoms can be fatal, the disease frequently goes into remission, which can persist for years. Because lupus can be fatal, it's critical to seek medical help rather than trying to treat it on your own. However, you can talk to your doctor about how to reduce your symptoms with natural therapies.

Anti-inflammatory herbs like turmeric and ginger, though not particularly investigated in lupus patients, may be especially beneficial for lupus patients who suffer from arthritic problems. Omega-3 fatty acids, which have been shown to reduce inflammation, have been proven to ameliorate symptoms in lupus patients in multiple trials. Oily fish (such as sardines and salmon) and flaxseeds, as well as a daily omega-3 supplementation, can help you get more omega-3 (Carvalho & Morales, 2010).

Alternative medicine for the management of lupus is too early to prescribe due to a lack of supporting research. Supplements haven't been thoroughly studied for safety, and because dietary supplements are largely uncontrolled, the content of certain items may differ from that stated on the label. Also keep in mind that supplemental safety has not been demonstrated in pregnant or breastfeeding women, children, or those who have medical conditions or are taking drugs.

Superoxide Dismutase

All live cells contain the enzyme superoxide dismutase. An enzyme is a substance that helps the body speed up chemical reactions. Cows are sometimes used to extract superoxide dismutase, which is utilized in medicine. Superoxide dismutase is a supplement that can help you get rid of wrinkles, regenerate tissue, and live longer. There is no indication, however, that superoxide dismutase products taken orally are metabolized by the body.

Superoxide dismutase is given as a shot to treat swelling and pain (inflammation) caused by arthritis, physical injury, and autoimmune diseases; interstitial cystitis, a kidney disorder; gout; paraquat exposure; cancer; and lung issues in neonates. Superoxide dismutase is also given as a shot to improve radiation treatment tolerance, improve rejection rates in kidney transplants, and reduce cardiac damage from heart attacks. For the treatment of corneal ulcers, a sterile solution containing superoxide dismutase is occasionally given directly to the eyes (Bonini et al., 2018).

Superoxide dismutase is an agent that aids in the decomposition of potentially damaging oxygen molecules in cells, reducing tissue damage. It's being studied to determine whether it can help with diseases where oxygen molecules are thought to play a function. In research investigations, injectable (shot) versions of superoxide dismutase are safe. Because certain superoxide dismutase products come from animals, there's a risk of contamination from sick or diseased animals. Although there have been no accounts of people becoming ill after using superoxide dismutase products derived from animals, it's recommended to stay away from them until more information becomes available.

The usage of superoxide dismutase during pregnancy and breastfeeding is not well understood. To be on the safe side, avoid using it. Superoxide dismutase is given as a shot to patients with certain bladder infections (interstitial cystitis), osteoarthritis, rheumatoid arthritis (RA), and lung damage that can occur in preterm newborns who have been given oxygen to assist them to survival.

Selective Serotonin Reuptake Inhibitor Sup Suppository

SSRIs work by improving the function of emotion-regulating nerve cells in the brain. Signals are used to transfer information between your brain cells. Neurotransmitters are the chemical messengers that carry these messages. One type of neurotransmitter is serotonin. When these brain cells (called neurons) communicate messages to one another, a small amount of a neurotransmitter is released to ensure that the message is delivered. They must then return the neurotransmitter they released before sending the next message. Reuptake is the term for the process of replenishing a neurotransmitter. The parts of the brain that regulate mood and convey messages utilizing serotonin may not work properly in depression. By inhibiting serotonin reuptake, SSRIs help to increase serotonin availability. This permits serotonin to build up between neurons, allowing for proper message transmission. Because they target serotonin, they're termed selective serotonin reuptake inhibitors.

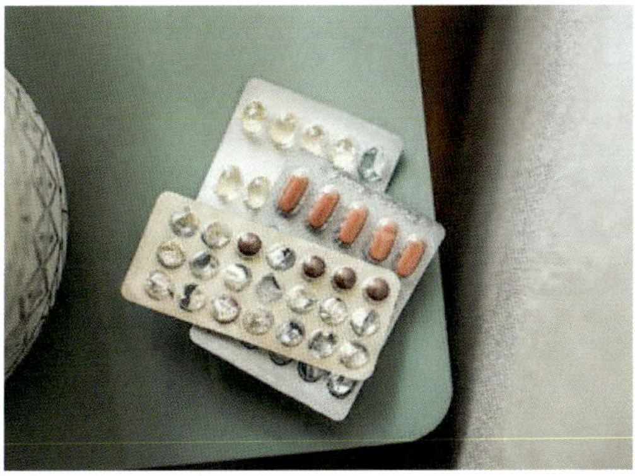

Figure 85: SSRIs are antidepressants that work by boosting serotonin levels in the brain. Serotonin is a chemical messenger (neurotransmitter) that transmits impulses between nerve cells in the brain (neurons). SSRIs prevent serotonin from being reabsorbed (taken back up) into neurons. This increases the amount of serotonin accessible, which improves message transmission between neurons. The selective nature of SSRIs comes from the fact that they only impact serotonin and not other neurotransmitters.

Source: https://www.medicalnewstoday.com/articles/6-natural-antidepressants

Most people who take SSRI antidepressants don't suffer serious side effects, but every medical treatment comes with some risk. These antidepressants may have the following adverse effects: Lethargy, Sleeplessness, Headaches, Rash, Blurred Vision Mouth dryness, Nervousness or agitation, Dizziness, joint or muscle pain If you have an upset stomach, vomiting, or diarrhea, Sexual drive is lessened, and Erection or ejaculation issues.

When using SSRIs, some persons, particularly children and young adults, may be more likely to experience suicidal thoughts. When compared to the results of taking a placebo, studies reveal that when taking any type of antidepressant, including an SSRI, the chances of having suicidal thoughts doubled — from 1 percent to 2 percent and 2 percent to 4 percent (Braun, 2005). Call 911 if you have thoughts of harming yourself while on an SSRI. There are also serious safety concerns with SSRIs to consider. If your system collects too much serotonin, you can get serotonin syndrome, which is an uncommon illness. This is particularly common when two separate serotonin-boosting drugs are taken.

Some prescription and over-the-counter medications, as well as herbs and vitamins, can interact dangerously with SSRIs. Make sure to tell your doctor about all of the medications and supplements you're taking before starting on an SSRI. Because all SSRIs act in the same way, the side effects are usually the same no matter which one you take. However, because each SSRI has a different chemical makeup, it's likely that if you're experiencing adverse effects from one, you'll have fewer or none if you switch to another. While some people experience adverse effects, others do not, and most side effects subside within a few weeks of medication. When it comes to noticing improvements from SSRIs, everyone is different. However, after 4 to 6 weeks of treatment, most people begin to notice positive ef*fects. It may take several months for the drug to take full effect.*

T

Tamarindus Indica

Also called Tamarinier, Tamarinier d'Inde, Tintiri, Imlee, Imli, Tamarin, and Tamarindo, Tamarind is a species of tree. The fruit, which is half dried, is used to create medication. Tamarind is used to treat constipation, liver and gallbladder problems, and stomach difficulties. Colds and fevers are also treated with it. Tamarind is occasionally used to relieve nausea caused by pregnancy. It is used to treat intestinal worms in youngsters. A thick paste made from tamarind seeds is sometimes used as a cast for shattered bones.

Eye drops containing tamarind seed extract are used to treat dry eyes. Tamarind is used to flavor meals and beverages. It's also a popular ingredient in Asian chutneys and curries. Tamarind includes chemicals that may have laxative properties as well as anti-fungal and antibacterial action. Because it includes a molecule that is similar to mucin present in the eye, researchers are looking into tamarind as a possible cure for dry eyes. Mucin protects and moistens the cornea's surface (Zhou et al., 2014).

Tamarind is a healthy fruit. When used in food levels, it is most likely safe. There isn't enough data to determine whether tamarind is safe to use as a medicine. There isn't enough trustworthy information on the safety of tamarind consumption while pregnant or nursing. Using in quantities greater than those found in foods should be avoided until further information becomes available.

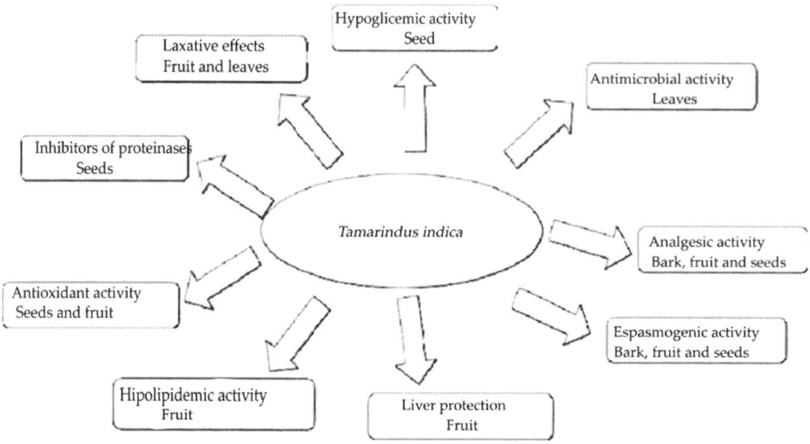

Figure 86: Tamarind leaf extracts have anti-oxidant properties in the liver and are used in cardiac and blood sugar-lowering medications. Young leaves can be used in rheumatism fomentation, administered to sores and wounds, or applied as a poultice to reduce swelling and relieve discomfort in joints. The leaves are used to make a sweetened decoction that can help with throat infections, coughs, fevers, and even intestinal worms.

Source: https://www.semanticscholar.org/paper/Tamarindus-indica-L.-A-plant-with-multiple-purposes-Menezes-Trevisan/f925327f31e4eb99c8ab3fd-96cee18629f79adc2

Tamarind has the potential to reduce blood sugar levels. It has been suggested that it may interfere with blood sugar regulation. If you have diabetes and consume tamarind, keep a tight eye on your blood sugar levels. It's possible that diabetes drugs will need to be dosed differently. Tamarind has the potential to reduce blood sugar levels. It's possible that it'll mess with blood sugar regulation during and after surgery. At least two weeks before a scheduled surgery, stop using tamarind.

The right amount of tamarind depends on a variety of factors, including the user's age, health, and other circumstances. There is currently insufficient scientific evidence to define a suitable dosing range for tamarind. Keep in mind that natural products aren't always safe, and that dosages are crucial. T. indica is a low-cost, easily accessible plant (Yi et al., 2016). Essential amino acids, phytochemical, and vitamins abound in it. It has a long list of well-known health advantages in traditional medicine. It could be employed in evidence-based treatment in a variety of health disorders with the help of current technology. More research into this plant's potential antioxidant and anti-inflammatory qualities, which could aid in the treatment of a variety of ailments, is required.

Tectona Grandis (Teak)

Common Teak (*Tectona grandis*) is a tropical hardwood tree that belongs to the Lamiaceae family of flowering plants, which also includes fragrant Premna, as well as herbs like motherwort, bugle, lemon balm, lavender, thyme, savory, oregano, holy basil, basil, marjoram, self-heal, wall germander, cat nip, ground ivy, Jupiter's sage, and hys Burmese teak, Central Province teak, and Nagpur teak are all names for different types of teak. The plant is native to south and Southeast Asia, including India, Sri Lanka, Thailand, Malaysia, Indonesia, Myanmar, and Bangladesh, although it has become naturalized and cultivated in many Caribbean and African countries.

The tree is currently widely planted in tropical and subtropical places around the world, not just for decorative purposes but also in commercial timber plantations. This tree's wood is a valuable tropical wood. When freshly milled, teak wood has a leather-like aroma and is prized for its toughness and water resistance. Boats, exterior construction, veneer, furniture, turnings, turnings, and other tiny wood tasks all employ the wood. Teak wood is tough, long-lasting, fine-grained, and resistant to rot when exposed to dampness, as well as having an appealing rich brown hue.

Teak leaves are cooling in nature, therefore they can help to reduce skin inflammation by acting as an anti-inflammatory agent. The leaves can be decocted or crushed to make herbal remedies for treating skin conditions. Leaves can also aid in the treatment of pruritus, a skin ailment characterized by acute itching. Leucoderma is a cutaneous disorder characterized by localized loss of skin pigmentation that can develop following a sequence of inflammatory skin conditions, post-dermabrasion, wounds, or intralesional steroid injections. Because leucoderma is mostly a cosmetic issue, it is neither communicable nor infectious (Van Jaarsveld, 2013).

Scabies can be treated using oil produced from teak blossoms. Scabies is a skin illness spread by the itch mite that causes itching and little raised red spots on the skin. Leprosy can be treated with bark. It's important to remember that leprosy is a contagious disease that causes discoloration and lumps on the skin and damages the skin, nerves, and mucous membranes. Leprosy can cause irreversible deformity and abnormalities in extreme cases.

Teak leaf extract is intended to help people overcome anemia, according to various research conducted by Togo scientists. Teak leaf extract was found to raise the concentrations of hemoglobin, platelets, hematocrit, and erythrocytes levels in young platelets when tested on mice. Teak flower oil can be used to promote hair development by applying it to the hair. Teak seeds are also traditionally produced and used as a hair tonic in the Indian system of medicine, according to another study. The effect of a petroleum ether extract of teak seeds on hair growth in albino mice was studied. Topically administered extracts of 5.00 percent and 10.00 percent in a simple ointment base were applied to albino mice's shaved denuded skin.

Terminalia Chebula

Also called *Terminalia bellirica, Terminalia chebula, Terminalia chebulic*, Thuulo harro Tropical Almond, Vibhitaki, Abhaya, Amandier Indien, Amandier Tropical, Arale, Arjan des Indes, Arjuna, Axjun Argun, Badamier, and Terminalia, chebula, Terminalia is a kind of plant. Terminalia species are utilized in medicine in three different ways. *Terminalia arjuna*, *Terminalia bellerica*, and *Terminalia chebula* are the three species. *Terminalia arjuna* has been used in ancient Ayurvedic medicine to balance the three "humors" of kapha, pitta, and vata. Asthma, bile duct diseases, scorpion stings, and poisonings have all been treated with it. For more than 3000 years, the bark of Terminalia arjuna has been utilized in India, particularly as a heart treatment. In the seventh century A.D., an Indian physician named Vagbhata is credited with being the first to employ this substance for heart ailments. Since the 1930s, researchers have been studying Terminalia, but the results have been varied (Umadevi et al., 2012).

People nowadays, however, employ *Terminalia arjuna* for heart and blood vessel diseases (cardiovascular illness), such as heart disease and accompanying chest pain, high blood pressure, and high cholesterol. It's also used as a "water pill," as well as treat earaches, diarrhea, sexually transmitted diseases (STDs), urinary tract illnesses, and to boost sexual desire.

Both *Terminalia bellerica* and *Terminalia chebula* are used to treat excessive cholesterol as well as digestive issues such as diarrhea and constipation, as well as indigestion. They've also been employed in the treatment of HIV infection. *Terminalia bellerica* is utilized to treat respiratory diseases such as respiratory tract infections, cough, and sore throat, as well as to safeguard the liver. Dysentery is treated by *Terminalia chebula*. For sore eyes, *Terminalia bellerica* and *Terminalia chebula* are used as an ointment. *Terminalia chebula* can also be used as a mouthwash or gargle.

Figure 87: *Terminalia chebula* (Combretaceae) is commonly used in Indian and Iranian traditional medicine to cure ailments such as dementia, constipation, and diabetes. The fruit of this tree, known as halileh or halilaj in Iranian traditional medicine (ITM), is utilized to produce cures.

Source: https://www.planetayurveda.com/library/haritaki-terminalia-chebula/

Terminalia chebula is used intravaginally as a douche to treat vaginal infections. *Terminalia bellerica* has been employed as a "health-harmonizer" in traditional Ayurvedic medicine, together with *Terminalia chebula* and *Emblica officinalis*. This mixture is also used to prevent cardiac tissue death and decrease cholesterol. Terminalia includes chemicals that aid in heart stimulation. It may also benefit the heart by decreasing blood pressure and cholesterol.

When ingested orally for 3 months or less, *Terminalia arjuna* is POSSIBLY SAFE. However, it should not be used without medical supervision. It could affect your heart. The safety of *Terminalia bellerica* and *Terminalia chebula* is not well understood. It's preferable to wait till more information is available before using it (Thakur, 2011). Along with standard therapy, 500 mg powdered bark of *Terminalia arjuna* every 8 hours is used to relieve chest pain after a heart attack. 500 mg powdered bark of *Terminalia arjuna* every 8 hours for congestive heart failure.

Tribulus Terrestis

Also called Tribulis, *Tribulis Terrestris, Tribulus terrestris*. Abrojo, Abrojos, Al-Gutub, Baijili, Espigón, Épine du Diable, Escarbot, German Tribulus Terrestris, Goathead, and Tribulus terrestis, Tribulus is a crop that grows spine-encrusted fruit. Tribulus is also recognized as a puncture vine because its spines are so sharp that they can deflate bicycle tires, according to legend. The fruits, leaves, and roots are used as medicines for a variety of ailments. Tribulus is used to treat sexual difficulties, infertility, chest discomfort, enlarged prostate, and a variety of other ailments, although there is little scientific evidence to back up most of these claims.

Sexual issues can make it difficult to feel satisfied during sexual activity. Taking tribulus by mouth can help ladies with sexual dysfunction or poor sexual desire have a better sexual experience. In males with poor sexual drive, taking Tribulus by mouth can improve sexual gratification. In athletes, taking tribulus by mouth, alone or in combination with other herbs and supplements, did not appear to improve body composition or exercise performance. There is some interest in employing tribulus for other uses, but there is not enough trustworthy data to say whether it will be beneficial.

When used at doses of 750-1500 mg daily for up to 90 days, tribulus may be considered safe for most people. Side effects are normally minor and infrequent, but stomach pain, cramping, and diarrhea are possible. There isn't enough credible evidence to tell if tribulus is safe to take for more than 90 days or what the potential adverse effects are. It is probably not a good idea to eat tribulus' spine-covered fruit. The fruit has been linked to major pulmonary problems in some people. Adults have traditionally taken 750-1200 mg of tribulus by mouth every day for 12 weeks. To find out what dose is best for a certain disease, speak with a healthcare professional.

Traditional Ayurvedic Medicine

Ayurveda, often known as Ayurvedic medicine, is an ancient Indian medical system. Ayurvedic medicine is a well-organized method of traditional health care that is widely practiced in Asia, both preventative and curative. Ayurveda has a lengthy history, with origins in India possibly dating back over 3,000 years. It is still a popular method of health treatment in many parts of the Eastern world, particularly in India, where it is used exclusively or in combination with modern medicine by a considerable percentage of the population (Stanisavljević et al., 2007).

The Indian Medical Council was established by the Indian government in 1971 to maintain undergraduate and postgraduate education standards. It defines appropriate qualifications in Indian medicine and acknowledges Ayurveda, Unani, and Siddha as traditional practices. Efforts have been made to blend traditional Indian and Western medical practices. The majority of Ayurvedic practitioners operate in rural regions, serving around 500 million people in India alone. As a result, they are a key force in basic health care, and the Indian government values their training and deployment.

Ayurveda, like scientific medicine, has preventive and curative features. The preventative component emphasizes the importance of adhering to a stringent personal and social hygiene code, the specifics of which are dependent on individual, climatic, and environmental factors. The restorative measures include physical workouts, the use of herbal remedies, and Yoga. Herbal remedies, external treatments, physiotherapy, and food are all used in Ayurveda to treat ailments. Ayurveda believes that preventive and therapeutic approaches should be tailored to the individual needs of each patient.

Dhanvantari, the gods' physician in Hindu mythology, is said to have learned Ayurveda from Brahma. Its early notions were laid out in the Atharvaveda, a section of the Vedas (c. 2nd millennium BCE). Vedic medicine was practiced until around 800 BCE. The Vedas are replete with magical rituals for the treatment of ailments, as well as charms for the expulsion of demons who are thought to be the source of disease. Fever (takman), cough, dehydration, diarrhea, dropsy (generalized edema), abscesses, convulsions, tumors, and skin illnesses are the most common conditions described (including leprosy). Many plants are indicated for treatment.

Traditional Chinese Medicine

TCM (Traditional Chinese Medicine) is a discipline of Chinese traditional medicine. The majority of its remedies have been described as having no rational mode of action, and it has been labeled as "fraught with pseudoscience." Traditional Chinese medicine included a variety of sometimes contradictory health and therapeutic methods, folk traditions, literati theory and Confucian philosophy, herbal cures, food, nutrition, and fitness, medical specialties, and philosophies, among other things. Traditional methods were seen as backward and unscientific by Chinese cultural and political reformers in the early twentieth century.

Figure 88: A health service that has been used to prevent, detect, and treat illness for thousands of years. It is founded on the notion that qi (the body's vital energy) travels through the body's meridians (channels) and maintains a person's emotional, spiritual, intellectual, and physical well-being. Traditional Chinese medicine tries to rebalance and harmonize the body's natural opposing energies of yin and yang, which can impede qi and lead to disease.

Source: https://www.bridgeport.edu/academics/schools-colleges/acupuncture-institute/traditional-chinese-medicine-ms

Traditional practitioners then integrated many aspects of philosophy and procedure into what they dubbed "Chinese Medicine" (Zhongyi). The Chinese government supported the fusion of Western and Chinese medicine in the 1950s, and marketed Chinese medicine as affordable and popular during the Great Proletarian Cultural Revolution in the 1960s.

"The body's essential energy (chi or qi) circulates through channels called meridians, which include branches associated with internal organs and functions," according to TCM. Its understanding of the human body is primarily focused on functions (such as eating, breathing, and maintaining body temperature) rather than anatomical structures: Both functions are grouped together and then assigned to a major functional entity — for example, tissue feeding and moisture maintenance are considered related functions, and the entity responsible for these functions is xu (blood) (Seth, 2003).

TCM uses pseudoscientific notions of the body and disease, comparable to Mediterranean humoral theory. TCM's body model is described as being full of pseudoscience. Some instructors no longer believe in yin and yang or the concept of energy flow. Traditional Chinese ideas like qi, meridians, and acupuncture points have no histopathological or physiological basis, according to scientific research.

Within China, TCM has been a source of contention. When Chinese academic Zhang Gongyao released an article titled "Farewell to Traditional Chinese Medicine" in 2006, he sparked a national debate by saying that TCM was a pseudoscience that should be eliminated from public healthcare and academia. However, because of the potential for export earnings, the Chinese government declared TCM to be a science and continues to promote its growth.

Tetrahydrocannabinol

THC, or tetrahydrocannabinol, is the molecule in marijuana that causes the majority of its psychological impact. According to the National Institute on Drug Abuse, it works similarly to the body's endogenous cannabinoid compounds (NIDA). Cannabinoid receptors are abundant in parts of the brain that are involved in thinking, recollection, pleasure, coordinating, and perception of time. According to the National Institute on Drug Abuse, THC binds to these receptors and activates them, affecting a person's memory, enjoyment, movements, thinking, focus, coordination, and sensory and time perception.

THC is one of several chemicals identified in the resin released by marijuana plant glands. The sexual organs of the plant have more of these glands than any other part of the plant. Cannabinoids, which are chemicals specific to marijuana, are contained in this resin. As per the National Center for Biotechnology Information, one cannabinoid, CBD, is non-psychoactive and prevents the high related to THC.

According to NIDA, THC causes dopamine release in brain cells, resulting in bliss. It also affects how data is processed in the hippocampus, which is essential for the creation of new memories in the brain. THC can generate hallucinations, altered thinking, and delusions. The effects last roughly two hours on average and begin 10 to 30 minutes after administration. However, psychomotor impairments may persist after the apparent high has passed.

Marijuana's properties make it a popular drug. It is, in fact, one of the most widely used illicit narcotics on the planet. However, mental health advocates are concerned about these consequences. According to NIDA, THC can cause a relapse in schizophrenia symptoms. Impaired motor skills are another potential side effect of THC consumption (Saslis-Lagoudakis et al., 2014). According to the National Highway Traffic Safety Administration, marijuana can impair driving or other duties for up to three hours after intake, and it is the second-most prevalent psychoactive substance discovered in drivers behind alcohol. People who use medical marijuana are advised not to drive unless they have proven that they can tolerate it and perform motor functions safely. Marijuana use may lead to issues for younger people. Some of them may be long-term, including cognition, memory, and a decrease in IQ.

Marijuana has been utilized for medicinal purposes for over 3,000 years, as per the National Cancer Institute. More than half of the United States had allowed the use of medicinal marijuana as of early 2017. Several states have also made marijuana legal for recreational consumption. THC may be taken from marijuana or manufactured, as the FDA-approved medication dronabinol demonstrates. According to the US National Library of Medicine, dronabinol is used to cure or prevent nausea and vomiting associated with cancer medications, as well as to boost the appetites of persons with AIDS. It is a yellow resinous oil with a light yellow color. Other research is accumulating evidence that THC has numerous more medical benefits when used properly. According to a 2016 study on mice, THC may be able to boost memory when consumed in tiny quantities.

Because of overdose, edibles, or meals containing THC, have become a major issue in jurisdictions that have legalized marijuana. People typically eat a whole quantity of a cookie instead of the prescribed lower amount when eating edibles, which can lead to overdosing.

According to the North American Industrial Hemp Council, hemp is a kind of cannabis with a low quantity of THC, as low as 0.5 percent. Hemp is employed in both the industrial and medical fields. THC content in some strains might be as low as 0.3 percent by weight. In certain strains, THC accounts for 20% of the total weight in a sample. The usual THC percentage in marijuana is 1 to 5%; in hashish, it is 5 to 15%; and in hashish oil, it is 20 percent, according to the National Highway Traffic Safety Administration. THC level in recreational marijuana doses varies greatly, and the lower the THC amount, the more the user must consume to achieve the desired effects.

Tumor Necrosis Factor

One of several proteins with a wide range of pro-inflammatory properties capable of inducing necrosis (death) in tumor cells. TNF is an abbreviation for tumor necrosis factor. TNF is a multifunctional cytokine that influences lipid metabolism, insulin resistance, coagulation, and the activity of blood vessel endothelial cells. TNF-blocking drugs have been demonstrated to be effective in lowering inflammation in inflammatory disorders like rheumatoid arthritis and Crohn's disease.

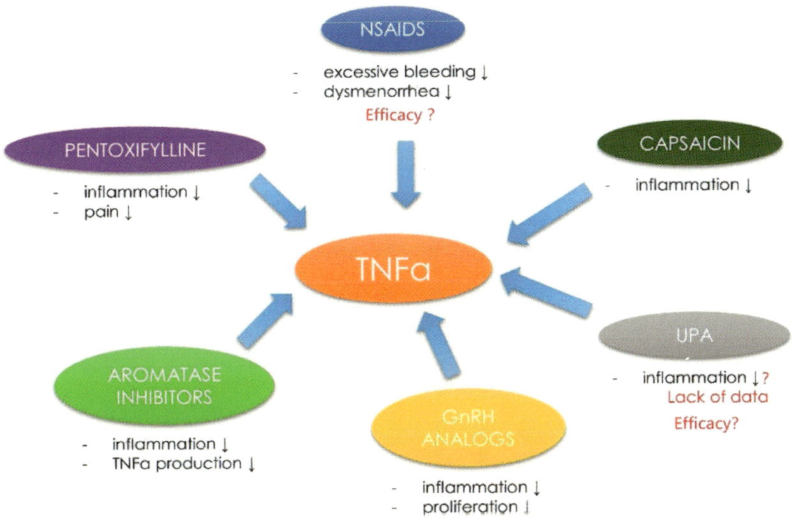

Figure 89: Tumor necrosis factor is not normally detectable in healthy people, but it is detected at high amounts in the blood and tissues in inflammatory and infectious disorders, and serum levels are linked to infection severity.

Source: https://healthjade.net/tumor-necrosis-factor/

TNF proteins circulate in the bloodstream to cause inflammation. They arrive at the desired location to start the inflammatory response. Excess TNF in the blood is deactivated by the body in healthy people, preventing it from causing inflammation. People can acquire an autoimmune disorder if this procedure does not perform properly. Autoimmune disorders are characterized by excessive inflammation, even when the body is not harmed. Rheumatoid and psoriatic arthritis are two examples.

High levels of TNF are not a cause for concern in healthy people. The body can control its immunological responses and prevent needless inflammation.

High amounts of TNF, on the other hand, can cause autoimmune disease flare-ups in persons who have it. Many autoimmune diseases have been related to high levels of TNF alpha in the blood, according to research. In such cases, the protein causes excessive inflammation, which results in sensations like pain. All of these illnesses are chronic and long-term, which means there is no cure (Nazim et al., 2020).

TNF is a protein that has a role in the inflammatory process. It is an important element of the immune system in healthy persons, assisting the body in mounting attacks against invading germs and viruses as well as healing injured tissues. Excess TNF in the blood can cause needless inflammation in persons with autoimmune disorders. This reaction can cause a variety of unpleasant symptoms. Long-term anti-TNF drugs are prescribed by doctors to treat a variety of inflammatory disorders, including arthritis and IBD. These treatments function by preventing excess TNF from being released into the bloodstream.

U

Uva Ursi

Also called Uva del Oso, Uva Ursi Extract, Uvae Ursi Folium, Arberry, Arbousier, Arbousier Traînant, Arbutus uva-ursi, Mountain Cranberry, Petit Buis, Ptarmigan Berry and common bearberry, Uva ursi is a flowering plant. Medicine is made from the leaves. The fruit is especially popular with bears, which explains its Latin name, "uva ursi," which translates to "bear's grape." Arctostaphylos uva-ursi is commonly referred to as uva ursi by most authorities. Some experts have also labeled the similar plants Arctostaphylos coactylis and Arctostaphylos adentricha as uva ursi.

Uva ursi is generally used to treat urinary tract infections, such as infections of the urethra, bladder, and kidney; urinary tract swelling (inflammation); excessive urination; pain when urinating; and urine containing excessive uric acid or even other acids. Uva ursi is also utilized to treat bronchitis, a lung ailment, and constipation. Compulsive bedwetting and painful urination can also be treated with a mixture of uva ursi, hops, and peppermint (Tuttolomondo et al., 2014).

Uva ursi is a plant that can help to reduce germs in the urine. It also has an astringent (drying) impact on the tissues and can minimize swelling (inflammation). When eaten by mouth for a short period, uva ursi is POSSIBLY SAFE for most individuals (for up to one month). Nausea, vomiting, stomach pain, and a greenish-brown colour of the urine are all possible side effects. When taken by mouth in excessive quantities or over a long period, however, uva ursi is POSSIBLY UNSAFE. It can result in liver damage, vision issues, breathing difficulties, convulsions, and death.

The right dose of uva ursi is determined by several factors, including the user's age, health, and other circumstances. There is currently insufficient scientific evidence to define an acceptable dosing range for uva ursi. Keep in mind that natural products are not always safe and that dosages are crucial. Before using, be sure to read the product label and consult your physician, pharmacist, or other healthcare experts.

Unicorn Root

Also called Plante Étoilée, Starwort, Unicornio Falso, Veratrum luteum, Alétris, Blazing Star, Chamaelire Doré, and Unicorn root, False unicorn is a type of plant. Medicine is made from the subterranean stem (rhizome) and root. False unicorn is used by women to treat ovarian cysts, menstruation issues, hormonal disorders, pregnancy sickness, and infertility. It is used by some women to restore hormonal balance after they have stopped taking birth control pills. False unicorn is also used to treat digestive issues and to increase urine flow, which helps to reduce water retention. Some people use it to get rid of worms in their intestines.

Chemicals found in false unicorns may promote the uterus and destroy intestinal worms. It may also cause an increase in urine output. There is insufficient evidence to determine whether a false unicorn is safe to use as a medication. Nausea and vomiting can occur at large doses. If you are expecting a child, do not use false unicorns. It may trigger uterine contractions, putting the pregnancy in jeopardy. There is not enough information on the safety of using false unicorns while breastfeeding. To be on the safe side, avoid using it. The stomach and intestines can be irritated by a false unicorns. If you have difficulties with your stomach or intestines, do not use it.

Figure 90: It's used to manage menstrual irregularities, menstrual cramping, and leucorrhoea, as well as a range of other conditions affecting the male and female reproductive organs. It can help with impotence, genito-urinary weakness or irritation, as well as liver and kidney disorders.

Source: https://www.healthbenefitstimes.com/false-unicorn/amp/

The impact of false unicorn could be similar to that of a water pill or "diuretic." False unicorn may impair the body's ability to eliminate lithium. This could cause major side effects by increasing the amount of lithium in the body. If you are on lithium, check with your doctor before using this product. It is possible that your lithium dosage needs to be adjusted. The correct dosage of false unicorn is determined by a number of factors, including the user's age, health, and other circumstances. There is currently insufficient scientific data to define an appropriate dosing range for false unicorn) (Teixidor-Toneu et al., 2018).

Upland Cotton

Also called Karpasa, Mian Hua Gen, Racine de Coton, Coton, Cotonnier, Cotton Plant, Cotton Root, Cotton Seed, Algodón, Algodón Americano, Algodón Cimarrón, and Algodonero, Cotton is a type of plant. The bark of the root is used to create medication. Cotton should not be confused with cottonseed extract (gossypol). Nerve pain, dysentery, diarrhea, headache, fever, nausea, and bleeding are all treated with cotton.

Cotton is used by women to treat menstruation problems and menopause symptoms. It's also used to induce labor and delivery, as well as to evacuate the afterbirth. Cotton is used by some women to help them produce more breast milk. Men sometimes use cotton for contraceptive methods, despite safety concerns. Cotton is also found in several vaginally administered birth control products. The cotton root bark has been shown to increase menstrual flow, induce childbirth, and function as a male contraceptive.

Cotton root bark formulations taken in levels found in foods are POSSIBLY SAFE for most persons when taken by mouth as a medication. Men who use cotton for birth control, on the other hand, should be aware that it may result in irreversible sterility. Cotton dosage is determined by a variety of factors, including the user's age, health, and other circumstances. There is currently insufficient scientific data to identify a suitable dosage range for cotton. Keep in mind that natural products are not always safe and that dosages are crucial.

Usnea

Also called Usnée Barbue, Usnée Fleurie, Usniate de Sodium, Usnic Acid, Woman's Long Hair, Acid Usnique, Barba de Capuchino, Barbe de Jupiter, and Tree Moss, Usnea is a lichen that can be found on trees. Lichens may appear to be solitary plants, but they are a mixture of fungus and algae that coexist for mutual benefit. Lichens grow in flat, bright patches. Usnea comes in a variety of colors, including yellowish, reddish, and black. Usnea's plant body is used to create medication (Silva et al., 2010).

Weight loss, pain relief, fever management, wound healing and making phlegm easier to cough out are all benefits of usnea. For sore mouths and throats, usnea can be applied directly to the skin. Usnea is not to be confused with oak moss (Evernia prunastri). Tree moss is a term used to describe both usnea and oak moss. Usnea contains chemicals that are supposed to help combat infection-causing bacteria. It may also help with inflammation, discomfort, and fever.

When administered on the skin, Usnea is POSSIBLY SAFE, while allergic responses can occur. Taking usnea by mouth could be dangerous. Usnea contains sodium usniate (usnic acid), which may cause liver damage.

Sodium usniate is a component of LipoKinetix, a weight-loss supplement with the brand name LipoKinetix. There have been multiple reports of liver injury in LipoKinetix users. Symptoms such as nausea, weakness, and exhaustion, as well as abdominal pain and skin yellowing, commonly appear 2 weeks to 3 months after commencing LipoKinetix. When you stop taking it, the symptoms go away. There was also one case of liver failure caused by ingesting pure usnic acid by mouth.

Figure 91: Usnea is a lichen that can be found on trees. Lichens may appear to be solitary plants, but they are actually a mixture of fungus and algae that coexist for mutual benefit. Lichens grow in flat, bright patches.

Source: https://www.healthbenefitstimes.com/usnea/amp/

The right dose of usnea is determined by a variety of factors, including the user's age, health, and other circumstances. There is currently insufficient scientific data to define an appropriate dosing range for usnea.

Uzara

Also called *Uzarae radix*, *Xysmalobium undulatum*, Uzara is a kind of plant. Medicine is made from the root. Uzara is used to treat diarrhea. Uzara contains chemicals that may cause the contents of the intestines to move more slowly. When taken by mouth for a short period, Uzara may be safe, although the adverse effects are unknown. Diarrhea that lasts more than a few days should be brought to your healthcare provider's notice. When taken by injection, Uzara is extremely dangerous and has resulted in death. The right dose of uzara is determined by many criteria, including the user's age, health, and other circumstances. There is currently insufficient scientific data to define a suitable dosing range for uzara.

If you have a cardiac issue, avoid using uzara. It contains substances that may aggravate your disease or cause treatment to fail. Potassium deficiency puts your heart's health at risk. Uzara use can drop potassium levels even further, increasing the risk of heart damage. Uzara has the potential to harm the heart. Some antibiotics may increase the amount of uzara absorbed by the body. Increasing the amount of uzara absorbed by the body may increase the drug's effects and side effects.

V

Valerian

The VALERIAN plant (*Valeriana officinalis*) has many nicknames. Roman soldiers used it to give them courage, for which they named it the "Soldier's Herb." It was thought to be sacred by the ancient Egyptians; Homer mentions its use, and Hippocrates prescribed it to calm nerves. As a part of many traditional folk medicine systems, valerian is believed to help with insomnia, anxiety disorders, headaches, respiratory problems and to relieve muscle spasms (Sharma M. J., 2010). The formula combines the benefits of a sativa and an indica into one strain. To reduce anxiety, it uses genetics from high THC strains such as Jack Herer and Haze, and then adds in some soothing properties normally associated with indica strains (Asnake et al., 2016). The result is an uplifting, happy high that won't leave you stuck on the couch. This native plant is one of the most fragrant flowers in North America, where it grows naturally in fields and throughout the country in a variety of habitats. It's even been planted as far away from its homeland as Africa to help control erosion. The other names that are used instead of valerian include: Garden Heliotrope, Garden Valerian, Grande Valerian, All-Heal, Amantilla, Baldrian, Baldrianwurzel, Indian Valerian, Mexican Valerian, Pacific Valerian, *Valeriana angustifolia, Valeriana edulis, Valeriana fauriei,*

Valeriana jatamansii, Valeriana officinalis, Valeriana angustifolia, Valeriana edulis, Valeriana fauriei, Valeriana jatamansii, Valeriana officinalis.

Figure 92: Valerian Plant.

Source: Photograph by NCCIH

It has a soothing, sophisticated blend of lavender, clary sage, ylang, bergamot and more. Scented with state-of-the-art aroma technology that releases fragrance at the first sign of stress. It has been used to treat insomnia and anxiety for centuries. In addition to being a great tasting drink, it is a natural stress reliever, muscle relaxant, headache relief, spasmolytic, and an orexin receptor agonist, helping you relax and unwind while offering a sense of calm for better sleep. The plant is used as a flavoring ingredient for foods and beverages, such as beer (e.g., Valerian Ale). As an herbal supplement, it is widely used in Europe, especially in Germany where is it used as a flavoring ingredient for the soft drink Valerianix and related products.

The consumption of valerian through the mouth is most likely safe, especially when used by adults in the amounts recommended by the doctor. It is required that one uses the plant or its by-products for the short term the safety of using valerian in the long term is unknown. Valerian is known to cause Headache, stomach upset uneasiness, excitability, heart disturbances and even insomnia for some individuals. For the case of feeling sluggish in the morning after taking Valerian, it is very few individuals experience such. It normally happens when one takes valerian in high doses (Sepúlveda-Cuellar et al., 2021). For others, it is the experience of vivid dreams and dry mouths. One precaution is that individuals are required not to drive after the consumption of valerian. When one wants to discontinue using Valerian, the best way is to reduce the dosage slowly over a week or even two before they can stop completely. There is not enough information regarding the use of valerian by pregnant women and breastfeeding ones. For children, it is probably safe for them to use within 4-8 weeks. In the case where one has been scheduled for surgery, then it is wise not to use valerian as the combined effects of valerian and the anesthesia can be harmful to the body.

In scientific research, the following dosage has been studied and accepted by doctors. 300 mg of valerian extract, 80 mg of passionflower extract, and 30 mg of hops extract before bedtime for up to two weeks. 374-500 mg of valerian extract plus 83.8-120 mg of hops extract before bedtime for 2-4 weeks. 120 mg of valerian extract, with 80 mg of lemon balm extract before bedtime for up to 30 days. Take valerian 30 minutes to 2 hours before bedtime. 400-900 mg valerian extract before bedtime for as long as 6 weeks.

Vitex Agnus-castus

Figure 93: Vitex Angus Castus plant.

Source: photograph by Alamy

Agnus-castus is a perennial deciduous shrub of the family Lamiaceae. It is a stunted, densely branched, procumbent, often spiny bush or suckering shrub, growing to 1-2 m tall by half as wide. *Vitex agnus-castus* is a plant that has been valued for centuries for its medicinal properties. The leaves are harvested at the end of summer and are preserved with natural alcohol to create Agnus Castus Leaf Extract, which is used in herbal medicine. This tantalizing liquid contains active constituents, including agnuside I and agnuside II. It is a flowering plant in the mallow family. Vitex Agnus Castus grows wells in most parts of the country, even in cold climates. Vitex agnus-castus has several interesting properties which make it potentially helpful for women who suffer from hormonal imbalances, menstrual irregularities and infertility.

A drawing of *Vitex agnus castus* (chaste tree) has been inscribed on an ancient Egyptian papyrus dating from 2,500 BC - 1,500 years before the biblical references. In the Middle Ages, the herb was recommended for witchcraft treatment and as a defense against the plague. It has been used for centuries to help address common symptoms associated with the transition into menopause, such as hot flashes or night sweats, mood swings, and irritability. In addition to its use for general menopausal symptoms, Vitex agnus-castus can also be used for treating PMS symptoms. It has astringent and antispasmodic properties and is used in herbal medicine primarily as a treatment for female reproductive issues, such as menstrual cramps, premenstrual syndrome, dysmenorrhea, menorrhagia, and bleeding from the uterus or vagina, vaginal discharge or leukorrhea, and breast pain associated with the menstrual cycle. The medicinal properties of Vitex plant have been proven by modern scientific methods. The most important bioactive components of Vitex agnus castus are flavonoids which are synthesized from several chemicals. These flavonoids include casticin, agnicin, castin and agnustilides (Salim et al., 2019).

Vitex is a generative plant, often used in medicines. It has no known side effects and it balances hormone levels. There is a study that claims that vitex seeds have the capability of repelling insects. Putting vitex with other plant extracts can protect individuals against head lice for up to seven hours. Furthermore, it impedes the reproduction of adult lice. There are also other benefits of vitex which include antibacterial and antifungal effects, reduced headaches, bone repair, reduced inflammation and the prevention of epilepsy. In most cases, this plant is considered to be safe. The side effects that have so far been reported are few and tend to be minor. Some of those side effects include; nausea, headache, mild skin rash, increased acne, mild skin rash and heavy menstrual flow. It is a recommendation that all pregnant women and nursing women avoid vitex given the fact that it has adverse effects on babies. There are those researchers that claim that vitex has interacted with birth control pills, antipsychotic drugs and birth control peels and thus they need to talk with their doctor before deciding to consume the same.

Veronica Anthelmintica

Beware the stare of this carnivorous species, *Veronica anthelmintica*. The yellow flowers are often visited by ants; once there, they fool them into believing that they are lost and looking for a nest. Thus, the insects take them to the nest only to drop them into a pit dug by the plant underground. The Veronica plant is one of the most popular and easiest to grow in the world. It's known for its long blooming period and ability to not need a lot of attention. Just set it and forget it! This is a wonderful gift for any plant lover, especially one who doesn't have a lot of time to tend to their garden (Schultz et al., 2020). The *Veronica anthelmintica* plant is an erect, herbaceous annual with a branching, purple stem and narrow leaf blade that is bronze on the upper side and silver or white on the underside. The plant grows between 1 and 2 feet in height and 0.4 to 0.6 inches in width. Leaves of this plant attach to its stems through short petioles. These plants are also known as speedwell, blue veronica, and scotch veronica. It has many common names including Rugweed, Carpetweed, Vinegarweed, and Mosquito Plant. It has one main growing system that can reach 9 feet long and often forms colonies that crowd out most other plants.

Figure 94: *Vernonia Anthelmintica* plant.

Source: African plants- a photo guide

They are very helpful plants given the fact that they repel parasites like pinworms and roundworms. Its crushed leaf used by the Apache people to treat gastrointestinal worms. It can be easily said that they are used for the relief of gastrointestinal problems, inflammation, and insect bites. The herb is very rich in camphor, which gives its leaves a strong smell. It can be used topically to reduce insect irritation, but keep in mind that it will also hurt the insects (think giant mosquito or spider), so use with caution. The *Veronica anthelmintica* plant (also referred to as the Speedwell plant) is used primarily to repel nematodes. The leaves of this plant contain coumarin, which exhibits anti-parasitic properties. Very few people know that *Veronica anthelmintica* is a plant that has been used for centuries to treat a wide range of medical conditions from asthma to hepatitis. The beneficial properties of this plant have been known since ancient times when Hippocrates recommended it to treat tapeworms and Galenus prescribed it to cure epilepsy. In addition, the plant has been traditionally used to treat scabies, indigestion, bladder ailments, scurvy and many others.

The common wormwood found in Europe and northern Africa produces a very powerful herb. The oil of the veronica plant is a powerful anthelmintic and one of the main ingredients in "wormwood wine" based on traditional usage for intestinal worms. The leaves have been used as a remedy for colds. The plant, its leaves and flowers contain several chemical compounds such as flavonoids, flavanones, flavones, phenolic acids and carotenoids (Saslis-Lagoudakis et al., 2011). All parts of the plant are utilized in herbal medicine, but the most important therapeutic components are considered to be the thujone-rich essential oil known as veronin, tannins, flavonoids, triterpenic acids and phenolic compounds. Traditionally Veronica anthelmintica has been used in folk medicine for its antihelmintic, analgesic, anti-inflammatory, diuretic, antiasthmatic, detoxifier, astringent and vulnerary actions. Its fruits showed antipyretic activity in acute treatment of children with influenza-like diseases. The presence of the following coumarin derivatives was reported: scopoletin (3β-(6-Cand 6-Gamyl)-2,4-coumaroylpiperidine), scopolin (6-(6-Candi 6-Gamyl)-2. The Veronica root is traditionally used for the treatment of respiratory diseases, urinary diseases, gastrointestinal disorders and fevers by indigenous populations. Furthermore, it has also been used to treat hemorrhoids, malaria, ear pain and rheumatism.

The stems of the plant are prostrate or ascending, sparingly, to filiform, with numerous leaves, which are always green and covered by a fine greyish-white mealy powdery coat. The leaves are alternate, lanceolate to linear, sessile at the base, oblanceolate or linear-lanceolate, 4–12 mm long and 1–3 mm broad.

There is little research that has been done on the product and as such, there are no specified dosage guidelines on how the product should be ingested. In most countries, the quantities that have been agreed upon on the consumption are mostly below 0.5mg/kg. those individuals that have a number of certain conditions are required to avoid wormwood at all costs. Those that are pregnant, can have a miscarriage. Those that are breastfeeding need not use the same because of a lack of safety information. For the case of those that have epilepsy, the consumption of this product can lead to seizures. Intestinal bleeding may take place when a person with heart disease consumes the product. The plant is also toxic to the kidneys and thus needs not to use as it may cause kidney failure. Those that have certain allergies of the Asteraceae family need to refrain from using the same.

Vigna Sinensis

Vigna sinensis is a plant species in the legume family, Fabaceae. It is a climbing plant with pinnate-leafleted vines up to 10 meters long, found widely across tropical Africa and Asia, and naturalized in Hawaii and the Antilles. In its native range, it is one of the primary food crops, cultivated for both its seeds and its young pods, which are used for human consumption. There are many cultivars grown for specific uses. They are commonly defined as the common mung bean, green gram,[1] or hare. It is a plant grown primarily for its edible seed. The seeds are used whole or split, boiled or fried in a variety of cuisines worldwide. *Vigna sinensis*, is the most widely cultivated species of the Vigna genus of flowering plants in the pea family Fabaceae and subfamily Faboideae. It is a small, erect perennial vine growing to 0.5–2 m tall, with slender stems. The leaves are alternate, and pinnate with two to four pairs of leaflets and a tendril. The flowers are pea-like, produced in racemes 2–7 cm long. The common bean is a very widely grown crop plant that has been cultivated for at least 7500 years. Its many varieties include green beans, wax (Reinaldo et al., 2020).

The plant is cultivated mainly in China (90% of world production), India, Bangladesh, Vietnam, Thailand, Indonesia, Philippines, Korea and the USA. It is an annual herbaceous plant growing to 40–100 cm tall. The leaves are 10–25 cm long and 5–13 cm broad, palmately lobed with five lobes. The flowers are 1–2 cm in diameter, with five white petals. It comprises about 60 species of annual and perennial herbaceous vines, climbers and shrublets. These species are native to Africa, Madagascar, southwestern Asia and Polynesia. The wild species are often used as food plants by the larvae of some Lepidoptera species. The Vigna be Tai (black-eyed pea) is one of the most diverse cultivated legumes with great potential for improving food security, and enhancing nutrition and health among millions of people in Africa.

Figure 95: Cowpea Vigna sinensis.

Source: photograph by ZHONG WEI Horticultural Products Company

Vigna sinensis seeds are harvested from mature soybeans and can be used as a substitute for traditional soybean products such as tofu and soy milk. It is high in fiber and very low in fat, making it a heart-healthy alternative to foods such as tofu. *Vigna sinensis* seeds are full of beneficial minerals including iron and calcium. Most importantly, these seeds provide high levels of isoflavones, which protect the heart and blood. It has been used as human food and traditional Chinese medicine for many centuries. It was domesticated independently in the Old-World tropics.

The major limiting factor of consuming this product is the fact that it has poor digestibility. This is because sulphur is present in the product which shows that the product has anti-nutritional factors. They contain several types of phenolic compounds which include proanthocyanidins, cyanogenic glucosides, oxalic acid, phytic acid, tanins and haemagglutinins. The consumption of the product has little information attached to the same concerning the correct dosage for the same. In case of any side effects, it is advisable to consume in small quantities and with the instruction of your physician.

Vitex Negundo

The *Vitex negundo* is a member of the Verbenaceae family, subfamily Lamiidoideae sub-tribe Verbenoideae. It is commonly known as Chastetree, Vitex, Chaste tree, Wild Lilac. It is a plant with large, oval-shape leaves and small, fragrant flowers and is a native of Java and the Philippines. Used as a medicinal herb for treating back pain, hypogonadism, and menopausal symptoms in Japan. Vitex is a tropical shrub with medicinal properties including contraceptive and anti-infertility properties (Rawat et al., 2020. Vitex has been used to promote female fertility and has been studied for its use in men. This attractive shrub grows to about 12 feet tall and has long, arching branches and smooth, green leaves covered with fine hairs. It bears clusters of bright yellow flowers in late summer and autumn. Vernacular names include love-looking glass, love-looking glass bush, heart leaf, and eye bright. Vitex negundo grows 30 feet high. It has edible fruit that is used to make jams, jellies, brown sugar, and other products. This tree is popular in areas where the soils are poor. It has been widely used as a medicinal plant by the Aztecs and other indigenous cultures for centuries and shows potential as a treatment for various medical conditions.

Figure 96: vitex negundo.

Source: photograph by useful tropical plants

Vitex Negundo is a medicinal plant used to reduce vaginal itching and helps women with menstrual disorders, irregular monthly cycle, pain during bleeding, and infertility. The fruit of this plant is highly medicinal and an expectorant, relieving sore throats and distressing coughs. Another name for the fruit is the False Ash Berry. In Vietnam, the fruits have been used to ease liver weakness and joint pain caused by rheumatoid arthritis. The fruits, roots, and branches have been used traditionally in Korea for their diuretic properties. In Japan, the bark has been used to treat diarrhea, dysentery, cholera, gonorrhea, and other gastrointestinal diseases. In Ayurveda, the plant has been used for colds, epilepsy and asthma. Asia often uses the plant as ornamentals. The leaves, stems, and fruits are used for flavoring food, especially soups and teas. It was said to have been used by Roman Vestal Virgins to maintain their vows of celibacy.

W

Walnuts

The walnut tree, probably Juglans regia (English walnut) is any of various trees of the genus Juglans (English h walnut; Persian h izgan; French h noix). The walnut is native to north-west Iberia and adjoining parts of France and Italy; it has been widely cultivated in England since Roman times. It is an economically important tree, widely grown for its produce, consisting of the nut itself, which is edible, the hard-shelled fruit or drupe that contains it, and the leaves that are used as food for livestock. Walnuts are a member of the hickory nut family and feature a meaty center that is rich in vitamin E and a myriad of other nutrients (Peng et al., 2017). Heart-healthy walnuts are grown in California's San Joaquin Valley and are dried slowly for a toasty flavor experience. Those walnuts have long been treasured for their health benefits when used as part of an active lifestyle. Walnuts are a delicious tree nuts that can be eaten whole or incorporated into recipes. The shells are used to make furniture and the nuts are harvested for food. Planting walnut trees create shade which reduces evaporation making it easier to keep our lakes full.

Figure 97: walnut tree.

Source: Walnut tree nursery Turkey

There are those described as Blue Diamond walnuts that originated in California and grow as they do in the golden state—in trees. Planted in 1882 and dedicated to Charles F. Shinn (who owned a large walnut orchard at that time), the grove is over 100 years old and is cared for with the belief that if you take care of them, they will take care of you. The reasons for loving walnuts are limitless including the fact that they are cholesterol and fat-free and they boast a high level of antioxidants and polyphenols, vitamins, minerals and plant protein. It is no doubt that Walnut trees are grown for sweet meat. If your only experience with walnuts is the less than tasteful supermarket harvest, you will be surprised by just how good fresh walnuts can be. That is because the remarkable flavour of this fruit is closely tied to its high oil content. To intensify this flavour, one needs to roast them in-shell. Then one should crack them so they can enjoy their full aroma and taste at their best.

There are those described as the black walnut tree which can live up to 600 years, yet its nuts—walnuts -with their similar name and rich taste, are among the most loved food around. A source of omega-3 fatty acids, this tree's nuts are also high in antioxidants and vitamins. Different names are used in describing walnuts which include dark-brown walnut, Juglandaceae (Latin), black walnut, walnut tree of the genus Juglans, mahogany, reddish-brown walnut and the English walnut (Ningthoujamet et al., 2012).

In the United States for culinary, health, and medicinal purposes long before colonial times. The Native Americans commonly referred to them as "horse chestnuts" because when cracked open, they look like a horse's heart. Walnuts are a popular nut. They have been found to help promote dental health because having a few walnut kernels after a meal can help minimize tooth decay. In addition, walnuts have been shown to lower cholesterol and promote heart health. They also provide beneficial amounts of omega-3 fatty acids and antioxidant vitamins and minerals.

Walnuts are medicinal plants. The walnut tree is a medicinal plant that helps prevent headache and nausea, but is also useful for gastrointestinal disorders, the coronary system, the genito-urinary tract, rheumatism, and skin diseases. It contains unsaturated fatty acids (oleic acid, linoleic acid). Walnuts have been used medicinally for thousands of years, dating back to the ancient civilizations of Egypt, Greece and Rome. Chopra suggests a walnut a day to shed a few pounds and keep the heart-healthy.

There needs to be a specific dose of consuming walnuts. This is given the fact that if high amounts are consumed, they may cause certain side effects. As such, it is required that one consumes the product in food amounts and also consult the physician on the right amount. For those that are pregnant, they need to consult with their doctor whenever they need to consume walnut supplements. Some of the side effects of walnuts include Bloating, Loose tools and also allergies.

Wasabi

Think of the wasabi plant as a lovely, peaceful little piece of nature you can grow in your house with the addition of tap water. The versatile, modern wasabia japonica grows so prolifically in tropical regions it's considered a nuisance weed. And that's just what you need when it comes to your food options. Wasabi is a member of the Nursery family. It grows as a perennial and is a flowering perennial. Being a member of the Nursery family, it falls under the umbrella of hardy perennials.it is a purple variety of the horseradish plant and grows vigorously and can grow up to 30 inches high. It has a taste that is hot and clean with a hint of fresh green color rather than traditional horseradish's overpowering flavor. In Japan, it is a frequently used ingredient in Japanese cuisine that has a sharp, hot flavor. Native to Japan, wasabi is a member of the Brassica family and is a relative of cabbage, broccoli and collard greens. The rhizome grows under shallow water in cool mountain streams.

Wasabi originates from Japan and is a horseradish-like root. It is grown similarly to horseradish, although it requires a lot more care and attention. The ginger plant is grown in a container where the roots are constantly sprayed with water. Wasabi plants are a natural accompaniment to sushi and sashimi. Some say the wasabi plant is a member of the cabbage family. Wasabi plants grow from 4 to 20 inches tall. Like true cabbages, these plants have dense, leafy growth at the base, with flowering stalks appearing from the middle. in Southeast Asia, it is known as bilis or seleri, in China (Sichuan), Korea, India (Kashmir), and other countries (Nadda et al., 2020). Wasabi (Eutrema wasabi) is a member of the horseradish family and not a true gourd. This amazing root is grown in beautiful greenhouses and can sometimes be found for sale at fish markets and other special events. The other names of Wasabi include Berstockrose, Cardamine pseudowasabi, Cochlearia wasabi, Eutrema japonicum, Eutrema wasabi, Gochunaengi, Japanese Horseradish, Japanese Wasabi, Japanischer Meerrettich, Korean Wasabi, Wasabia japonica, Wasabia japonica var. Korean, Wasabia koreana.

Figure 98: The wasabi plant.

Source: Britannica

This plant is known for its pungent odor when touched or bruised. Most is cultivated in Japan, China and Taiwan. Widely known as the spice in American sushi restaurants, it has many medical properties in China. Wasabi is a green plant widely used as a spice and as a medicine. Like many roots or bulb vegetables, it contains high levels of glucosinolates, substances that have been shown to have anti-carcinogenic properties. The root is very spicy with two major compounds naturally occurring in wasabi, namely allyl isothiocyanate and sinigrin. Allyl isothiocyanate causes the sensation of heat in the human nose when consumed, along with tingling and watering of the eyes. For adults, a teaspoonful of the plant is required. For the case of pregnant and lactating mothers, it is recommended that they consume the product with the instruction of their physician (Miara et al., 2018).

Watercress

The plant is a perennial herb with edible green leaves and small white flowers. Watercress is a healthy green vegetable that requires little effort to grow. Watercress thrives in cold water streams or partially submerged environments, particularly those rich in calcium from limestone or chalk. Watercress is native to Europe and western Asia where it can be found growing wild in clean flowing rivers and streams. The fleshy, edible leaves of this aquatic herb have a peppery flavor that resembles horseradish or mustard. It is a staple of home gardens and a favorite vegetable in many parts of the world. The young shoots are eaten as a spring vegetable, much like asparagus, and the roots can be eaten as a winter crop. Watercress is a hardy perennial growing up to a meter tall, with hairy stems and leaves. They are described as the Magic Plants of Maui, takes you on a wild food adventure to sample excellent-tasting plants straight from your garden. From delicious grapes to tasty tree ferns, each plant is probably growing nearby your home.

Figure 99: Watercress plant.

Source: Britannica

Watercress is a special plant. Its scientific name is _Nasturtium officinale_. It grows easily in ponds and shallow water, as long as it gets plenty of sunshine. Watercress is a very fast-growing plant that can attain harvestable size in about 20 days, depending on the season. It has a crunchy texture with a slight peppery taste similar to radish or arugula. This delicious green has been enjoyed for centuries thanks to its high vitamin content, calcium, iodine, iron and water content. Watercress grows especially well near running water or drawn up from wells of mildly acidic groundwater of at least neutral pH. As soon as the sun warms the earth, this small but tasty plant thrives on its sunlight and transforms into darkly-hued roots (Menendez-Baceta et al., 2015). These roots become the raw ingredient for the Fire Cured Virginia Watercress Collection. Watercress, as a semi-aquatic annual plant, is a very healthy fresh vegetable, often consumed as a green leafy salad vegetable. Like most members of the Brassicaceae family, it has a distinctive pungent flavor. Watercress is sold loose or bunched with roots and dirt removed, and either packed in a sealed plastic bag or wrapped in damp peat or plastic film. This plant is also known as nasturtium. Cresson Officinal, Indian Cress, Jal-Halim, Mizu-Garashi, Nasilord, Nasturtii Herba, Nasturtium officinale, Agriao, Berro, Berro di Agua, Berros, Brunnenkresse, Crescione di Fonte, Cresson, Oranda-Garashi, Radicula nasturtium, Rorippa nasturtium, Scurvy Grass, Cresson au Poulet, Cresson d'eau, Cresson de Fontaine, Cresson de Ruisseau, Sisymbrium nasturtium, Selada-Air, Tall Nasturtium, Wasserkresse, Waterkres.

The leaves and stems of the watercress plant acts as a healing food for stomach and chest problems. It is most often used in salads, sandwiches, soup, and is also an important ingredient for herbal tea. The highest concentration of the essential mineral iodine can be found in watercress, making it a surprisingly strong contender as a remedy for iodine deficiency. Considered a vegetable, watercress is similar to mustard and arugula. This healthy green has incredible detox power and is a whopping 20 percent protein. Called one of the world's most nutrient-dense foods, Watercress contains vitamins A and C, as well as folic acid. Watercress has long been thought to prevent tooth decay due to its high concentration of calcium, phosphorus, and fluoride.

When consumed by mouth, this plant is likely to be safe in food amounts. It is also considered to be safe if consumed in amounts used in medicine even if consumed through the mouth. This is normally in the short term.

However, if consumed in the long term and possibly in large amounts, this product can be unsafe and known to cause damage to the stomach. Even in medicinal amounts, this plant is likely to be unsafe for women that are pregnant or breastfeeding (Muchugi et al., 2006). It can lead to menstruation and can lead to miscarriages. As such, it is rather these individuals do not use the product. For the case of children under the age of four years, this plant is unsafe for them. The suggested dosing of the plant is: Dried Watercress: 4-6 g/day, Fresh Watercress: 20-30 g/day, Watercress Juice: 60-150 g/day. Furthermore, the consumption depends on the health of the user, their age and other conditions.

Wheatgrass

Wheatgrass is an easy-to-grow perennial rich in chlorophyll. One cup of wheatgrass has 18% of the daily value of fiber, 10 times more than spinach does! Wheatgrass juice lasts up to 14 days before it starts to lose enzymes and beneficial nutrients. You don't have to be a farmer to grow your wheatgrass. It is a fast-growing grass, becoming seeding size in as little as 45 days. This makes it a very easy plant to grow and maintain. The young wheatgrass shoots can be juiced just like other green leafy vegetables and may be grown indoors on a countertop during the winter months. The synonyms of the plant include Couch grass, Couch Grass, Cutch, Dog Grass, Dog-grass, Doggrass, Durfa Grass, Agropyre, Agropyron, Agropyron repens, Agropyron firmum, Blé en Herbe, Brote del Trigo, Quitch Grass, Scotch Quelch, Triticum, Triticum firmum, Triticum repens, Twitchgrass, Wheat Grass, Witch Grass, Elytrigia repens, Elymus repens, Graminis Rhizoma, Herbe de Blé, Quack Grass, Quackgrass.

Wheatgrass is a particularly nutritious form of grass, which is normally grown as a commercial crop to be ground into flour for use as a gluten-free whole wheat substitute. The edible portion of this plant consists of the young soft stems that emerge from the seed head of the grass after the grain has fully developed (Maruca et al., 2019). Wheatgrass contains high concentrations of chlorophyll and is rich in minerals, vitamins, enzymes, and plant hormones including alkaline ionized polysaccharides. It also contains more protein than any other plant on earth.

Figure 100: wheatgrass plant and juice.

Source: photograph by Pinterest

Wheatgrass is a medicinal plant that grows in a variety of climates. Wheatgrass juice contains vitamins, minerals, amino acids and chlorophyll. In addition, wheatgrass juice is beneficial to overall health as it can help to detoxify the liver and blood by eliminating excess cholesterol from the body. This is a type of grass with many medicinal benefits. This seed of the grass has been known to cleanse the body. It is considered to have cancer fighting properties, increase athletic endurance, assist digestion, detoxify the blood, slow down the aging process, and promote overall health (Martínez & Luján, 2011). It was first used by Pharaohs from ancient Egypt. Wheatgrass Juice Powder is crammed with nutrients, and rich in enzymes and chlorophyll – which helps promote blood health and greatly assists the body to cleanse and detoxify. With a mouth-watering fresh flavour and a uniquely pleasant aroma, this sensational juice powder offers an excellent source of minerals including calcium, potassium and magnesium, as well as vitamins A, C and E. Furthermore, it helps wih the digestion as it helps the body break down food and absorb nutrients. It also boosts a person metabolism and aid in weight loss.

The possible side effects of consuming the plant includes, nausea, headache, constipation, upset stomach, and fever. In most instances, these side effects occur when the product is newly introduced in the body but likely to fade away once the body has adjusted to it. It is recommended that one purchases wheatgrass from a trusted supplier. As a new consumer, it is advisable that they first take the product in small quantities as they gradually increase to meet the recommended dosage. In a liquid dose, the typical amount is from 1 to 4 ounces (oz.), or about 2 shots. For the usual powdered dose, it is about one teaspoon or 3 to 5 grams. After taking the product, one should drink water to reduce the side effects.

Whey Isolate/ Whey Protein Isolate

Whey is a plant-based protein product. Whey is made by separating the water-soluble portion of milk using a microfiltered process. This concentrate is then spray-dried into a powder form. Whey protein exists in two basic forms, whey protein concentrate and whey protein isolate. Ingredients: 100% Dairy Free Whey Protein Isolate from Grass-Fed Cows, Natural Flavors from Non-GMO Lactose Free Milk, Sea Salt. This whey is sourced from grass-fed cows in Ireland and Italy. On average, cows receive an abundant amount of pasture and high-quality crop residues in the form of beet pulp and sunflower stalks to supplement low protein grass. The result is a mild, sweet-tasting whey that contains a host of beneficial nutrients from grass silage (Mati & de Boer, 2011).

NutraPlanet's Whey is rich in naturally occurring, completely absorbed protein. It contains all 9 essential amino acids (including branched-chain amino acids L-leucine, L-isoleucine, and L-valine) in their optimal ratios for muscle development, repair and recovery. It's also high in glutamine and calcium caseinate, which is also high in glutamine, considered the #1 amino acid for muscle growth. Whey protein contains digest quickly into muscle tissue following exercise helping to promote recovery between workouts. Whey is a delicious and nutritious protein supplement designed to help build muscle and decrease recovery time after workouts. It has been called the "Bridge Between Meals" because it provides the extra protein that many of our bodies need but do not consume. As a dietary supplement, Whey protein is an excellent source of high-quality protein and BCAAs and is clinically proven to promote lean body mass and decrease recovery time after exercise. Whey protein is quickly absorbed by muscles without the use of regular digestion.

Whey is better than milk. Why? Because using Whey protein leads to more muscle, less fat and improved lean body mass. Using Whey leads to more strength, endurance, recovery and protein synthesis (protein building).

It leaves you more satisfied for longer. It has 5-6g of BCAA's per scoop which is 3x's the amount of your normal protein shake. Protein contributes to the growth and maintenance of muscles (Li et al., 2015). The other names of Whey include Fraction de Petit-Lait, Goat Milk Whey, Goat Whey, Isolat de Protéine de Lactosérum, Milk Protein, Milk Protein Isolate, Mineral Whey Concentrate, Bovine Whey Protein Concentrate, Concentré de Protéine de Petit-Lait Bovin, Whey Fraction, Whey Peptides, Whey Protein Concentrate, Whey Protein Hydrolysate, Whey Protein Isolate, WPC, WPI.

Whey is a medicinal product containing, in addition to lactose, also milk proteins (whey proteins 20% - 30%), fats (2% - 6%) and minerals (Calcium 4.5%, Iron 27mg/kg, Zinc 7.5mg/kg). The whey is produced during the process of production of cheese. It is used especially in infant nutrition for nutritional support in children with weakened digestive status or anorexia, or in children with increased needs of nutrients. It is not intended to substitute for a varied diet. It is obtained from the by-products of cheese production and contains alpha-lactalbumin and beta-lactoglobulin as its main protein components. Whey is used as a nutritional supplement; as an ingredient for medical food products (other than infant formulas); as an ingredient in pharmaceutical preparations; as a constituent of infant formulas; as a source of special dairy foods; and as a therapeutic substance or dietetic substance.

Not for use in children under 12 years of age, except under medical supervision. Keep out of reach of children to avoid unintentional consumption. Storage: Store in a cool dark place away from foodstuffs, once opened store in an airtight container in the fridge. Use within 4 weeks following opening. The daily protein needs normally varies on a number of factors such as age, gender, weight, and the level of physical activity (López-Malo et al., 2020). The amount of whey that can be consumed normally depends on the reason behind the consumption of the product. The possible side effects of this plant are that it can lead to increased bowel movement, bloating, reduced appetite, thirst, nausea, acne, tiredness and headache. This product is often used by adults in doses of 30g per mouth daily. It is recommended to use for up to 6 months. It is however important that one speaks with a healthcare provider considering the amounts to be used for different purposes.

White Oak

Figure 101: White oak tree.

Source: Wikipedia

The white oak is a large deciduous tree native to North America, Southern Europe, Central Europe, Western Asia, Northeastern Asia, China, Korea, Japan, and Siberia. It is one of the two most common oaks in the eastern United States. White Oak is a prominent hardwood species of the temperate regions of the Northern Hemisphere. It is found in Europe, Asia, North America, and to a lesser extent South America. It is an important timber used in construction due to its strength and shock resistance. The wood was widely used for bunkers on U.S Navy ships until the 1990s, when it was replaced by steel to reduce weight (Kujawska et al., 2017). Some botanists believe that all white oak groups should be considered species in light of genetic research. White oak will achieve a height above 100 feet and a diameter over 7 feet. As a result, it is a magnificent specimen for a landscape in a park or large commercial setting.

The white oak (Quercus alba) is scientifically classified as the only species in the section Leucobalanus native to North America. White oak is an angiosperm belonging to the beech family Fagaceae. The drop-shaped acorns are a distinguishing characteristic of this species, and also lend their name to the tree's vernacular names. Other common names include "southern red oak", "swamp white oak", "burr oak", and "Crawford oak". The sapwood is very white with red streaks, while the heartwood varies from nearly white to a light reddish-brown. The lumber is widely used for furniture, cooperage, flooring, railroad ties, mine timbers and other construction purposes. White Oak's natural beauty and rugged durability make it an excellent choice for landscape architects and designers. It is also well suited to survive in roadside planting locations where trees are subject to damage from vehicles (Kunwar et al., 2016). In the past, when building construction called for large beams, white oak was used quite often because its high strength-to-weight ratio made it the preferred choice when strength was critical. White Oak Leather comes from the same trees that produce White Oak lumber. Each hide is carefully selected by hand, inspected for veining and blemishes, then sanded to a fine finish. Each natural piece of leather is finished with our signature oil for lasting durability. Every hide undergoes an extensive grinding process that preserves all-natural markings. The result is smooth, finished leather that is easy to clean.

The White Oak Tree is a medicinal plant in the United States. The leaves, bark and acorns have been used for medicinal purposes. The white oak has an aroma like acidulous camphor. It stimulates circulation, corrects stomach problems and ulcers, and fights worms and rheumatism, but because of the large number of tannins it should be taken in small doses because it also acts as a diuretic and purgative after a time. As an acrid, astringent and highly aromatic medicinal herb, White Oak Bark was used to make tonics and teas to treat colds, fevers and other upper respiratory ailments. It has also been used as a gargle for sore throats and as a poultice for swollen glands and skin infections. White oak leaves can be made into an infusion to use as a drink to relieve pain, congestion and diarrhea. Don't take mugwort internally without consulting a professional first. It is also used in treating gallbladder problems, intestinal worms, stomach inflammation, and diarrhea, and for healing cuts and sores.

This plant might be safe to use if consumed 3-4 days in the case of diarrhea. However, it may lead to side effects such as stomach and intestinal problems, and kidney and liver damage. In most, it is considered to be safe for individuals that apply the same to the skin. However, it is required that it is used 3-4 weeks. When taken longer than that, it may cause skin damage (Kovach et al., 2009). It is required that those individuals with heart conditions not to use the product. For the case of pregnant and lactating women, little is known and as such is safe to consult with a doctor. The dosing of this product usually depends on a number of factors such as the user's age, health and several other conditions. As such, it is advisable to contact ones physician.

Wild Yam

This plant is cultivated for its succulent, tuberous rhizome. A rich, full-bodied herbaceous perennial root, harvested at 8-10 inches. Diosgenin is an alkaloid found in the wild yam or lowan. Wild yam has been used to help balance hormones, relieve menopause symptoms and help with sleep. Having antioxidant properties, wild yam helps to boost the body's immune system. Wild yam is a native plant found in some areas of the United States. Wild yam, a member of the Dioscoreacea family, is a long-lived perennial herb that grows in thick patches over a large part of the south United States. It has a significant root system and a strong taproot upon which it can sprout from when cut. The leaves are from 7 to 15 inches in length with thick blades that have serrated edges with curved bases. The root is a rich source of polysaccharides, which some think help to relieve symptoms associated with menopause (Karmakar et al., 2010). Wild Yam is a safe, potent, natural alternative to hormone replacement therapy. 11 herbs and wild yam extract help re-balance the levels of estrogen and progesterone in the body. Winni Winni's Wild Yam formula has been shown to balance hormones and reduce pain associated with PMS. The synonyms include American yam, Chinese yam, Colic root, Devil's bone, Four-leaf yam, Mexican yam, Rheumatism root, Shan Yao (in traditional Chinese medicine) or Yuma.

Figure 102: Wild Yam.

Source: photograph by HerbaZest

Wild yam seeds have been used by Native Americans since time immemorial for a variety of purposes, including childbirth and the prevention of cramps. This potent herb is believed to have several medicinal purposes. It is said to be an effective pain reliever, help with menopause symptoms, to help reduce the look of varicose veins, liver disorders, and kidney stones, and may even be able to fight cancer cells. Apart from being a valued medicinal herb, it is also a powerful culinary herb. In addition to its ease of cultivation, the fact that the root is used as well as the leaves give it an advantage over competing herbs. Wild yam root, or diosgenin, is an ingredient in many pharmaceuticals. Wild yam contains diosgenin, progesterone, puerarin, calabazin, including some unidentified steroid acids. Wild yam was found useful in the treatment of gynecological disorders that are associated with low levels of progesterone. The leaves are most often used in herbal medicine for their diuretic properties. Wild Yam contains natural sources of diosgenin, which is used to synthesize progesterone. Wild Yam is also the primary ingredient in bioidentical hormone replacement therapy (BHRT).

Wild yam contains diosgenin, which is used in the manufacture of most oral contraceptives. Studies also show that wild yam creme helps improve the health of the uterus, making it efficacious in treating problems like endometriosis or menopausal symptoms, such as lack of libido. Wild yam is an herbal blood purifier and is also used to treat intestinal disorders, spinal cord injuries, hemorrhoids, and cancerous tumors. The anti-microbial action of wild yam root may be due to the steroid content and there's evidence that the herb is beneficial for treating fungal infections like candidiasis and thrush (Kayani et al., 2015).

The plant is considered to be safe when taken orally or as a topical cream. The side effects may be mild including diarrhea, vomiting, stomach upset or nausea. These are most likely to occur when consumed in higher doses. Wild yam extract is a common cause of gynecomastia, a condition in which males develop enlarged breast glands due to a hormone imbalance. Little knowledge is known regarding the safety of consuming these drugs when one is pregnant or breastfeeding. With regards to the dosage, there are no certain guidelines concerning the best way to use the same. The wild yam supplements are normally sold in tablets or capsules ranging from 100-400 milligrams. Generally, they are safe to consume one or twice in a day, with or without food.

Wormwood

The botanical name for Wormwood is Artemisia absinthium. The name itself was used by the ancients to denote some fantastic celestial plant of great destination, some magic herb that grew only at the gate of Paradise. An unusual plant with a bitter taste. The Wormwood is a large shrub growing to 4–8 m tall. The leaves are bipinnate, with leaf segments 10–20 cm long and 2–3 cm broad. The flowers are small, pale yellow-green, produced in dense compound umbels. Takes on its usual dark green coloration when exposed to sunlight. This tender perennial plant has a long history of use. Made from wormwood and other botanicals, the drink called Absinthe was a favorite drink of artists, writers and other free thinkers around the turn of the last century. It is known as the bitter herb; the bitterness of Wormwood was thought to preserve one from death (Kassam et al., 2010). It is native to the Northern Hemisphere and has oblong, highly divided, feathery leaves that give off a faint odor when crushed. The plant has a taproot and produces flowers with yellow-green petals and an orange ring around the seed head at the end of long, thin stems. The plant has been traditionally used as an aromatic bitter herb, as a sedative, and as functional food (in small doses). The Wormwood plant was considered by the early alchemists to be the most prized substance on earth. It was said to have magical properties able to transform ordinary materials into precious metals. Chemically, wormwood is known for containing thujone, a plant-derived compound that has potent psychoactive effects in humans. This inspired the medieval legend of witches flying through the night on brooms made out of this herb. Used in the past to treat malaria and other fevers. In ancient times wormwood was used for many purposes including embalming, anointing, fumigation, insect repellant and flavoring. Many of these uses have been forgotten since more effective treatments have been developed. In the current society, individuals are normally seen to mix them with whiskey, tequila, blood orange juice, ginger beer or even soda water for an unforgettable experience.

Figure 103: Wormwood plant.

Source: Dr. Axe

Wormwood is a perennial herb that was once used in European herbal traditions to fight fatigue, stimulate the mind, ease gastrointestinal issues, and treat wounds. Wormwood's reputation as a cure-all stretches back thousands of years. The plant is filled with a white milky juice. It is bitter and possess a sharp biting taste, hence the name wormwood. The whole plant is full of a strong, pungent, bitter spirit essential oil. The young leaves are somewhat more delicate in flavor than the older ones. The juice from the leaves has traditionally been used in herbal medicine and wines, and the dried stems in herbalism (Jiang et al., 2019). The bitter aromatic leaves were also used as a diuretic and for digestive and respiratory disorders. The bitters are extracted from wormwood to make the spirit absinthe. It is believed that Wormwood has chemicals that can help stimulate dreams and nightmares. Its active constituents are sesquiterpene lactones, which are the product of the enzymatic transformation of nortriterpenoids via the action of the enzyme Artemisia lactone formate-lyase. This enzyme is encoded by coleopteran-specific NADPH-dependent oxidoreductases. Wormwood was known to be used for alleviating pain, fighting parasitic infections, boosting antioxidant properties and fighting inflammation.

Side effects of the product are normally seen in individuals with certain conditions. These individuals are therefore required not to use the wormwood. Such are those with, pregnancy, Breastfeeding and children, Epilepsy, heart disease, kidney problems and those that have allergic reactions to the plants belonging to the Asteraceae family. Due to a lack of research on the same, there is no specific dosage for the product. Given the fact that the plant has toxic effects, most governmental institutions around the world have placed restrictions on use. In case an individual is unsure concerning the way to use the product, it is best they speak to their medical provider.

Withania Somnifera

Figure 104: Withania somnifera.

Source: Photograph by Seeds Gallery Shop

Withania somnifera is a plant commonly called ashwagandha, Indian ginseng, or winter cherry (in West Bengal). It is a short herbaceous perennial plant in the family Solanaceae. It naturally grows in the Indian subcontinent and Afghanistan, but is cultivated widely elsewhere on a commercial scale. The withania plant is a small shrub with both white and dark purple flowers. *Withania somnifera* Linn. (Solanaceae/Nightshade family) is an herbaceous annual plant that grows between 50 cm to 100 cm tall. It is commonly known as Ashwagandha, Akeppali, Amishwakarma, Wintercherry, Indian ginseng, Ulva-Indica, Black Nightshade etc. When it comes to its use in ayurvedic medicine it is referred to as ashvagandha. The roots of ashvagandha are the most commonly used part for medicinal purposes. The Withania plant is named after the Greek Goddess of fertility and healing, Asclepius. The other names of the plant include Avarada, Ayurvedic Ginseng, Cerise d'Hiver, Clustered Wintercherry, Ghoda Asoda, Ginseng Ayurvédique, Ginseng Indien, Hayahvaya, Indian Ginseng, Sogade-Beru, Strychnos, Turangi-Ghanda, Vajigandha, Winter Cherry, Withania (Jima & Megersa, 2018).

For hundreds, maybe thousands of years, withania has been used as a medicinal plant in the Indian Ayurvedic system of medicine. The roots have been shown to have various therapeutic uses, including being effective in treating ulcerative colitis and mild anxiety. The roots of this plant are used in Ayurvedic medicine where it is regarded as a medicinal herb that helps rejuvenate the body's tissues and organs. It is said to be beneficial for treating conditions of the genitourinary tract, reproductive disorders, insomnia, neuromuscular diseases, inflammatory diseases, alimentary tract disorders, cardiovascular diseases, infection by dangerous microorganisms, convulsions, tumors, anemia, leucoderma etc. It is known for its role in stimulating the immune system and overall well-being and has been used to treat such conditions as asthma, arthritis, diabetes, multiple sclerosis, and nervous system disorders. The active constituents in Ashwagandha include steroidal lactones (withanolides), including within one and others; these give the root its distinctive smell.

Sava is a Sanskrit word, used to name the Withania somnifera, meaning "all-healer." It's a name we gave to this cultivated plant after observing its benefits on every organ of the body. Sava is a natural product that has been used for centuries by women to increase energy and stamina, improve bone health, promote mental alertness, reduce stress and fatigue and support healthy aging. Active components within Ashwagandha's root, leaf, stem, and fruit help support the body during stressful times. From boosting the immune system to maintaining healthy cortisol levels, Withania somnifera helps promote homeostasis and well-being.

When consumed through the mouth, the product is possibly safe, but only for up to three months. Its long-term safety is currently not known. In large doses, it has side effects which include, liver problems, stomach upsets, diarrhea and vomiting. It is on rare occasions that liver problems might occur. With regards to its application to the skin, there is not enough information that one can be able to rely on. It is likely unsafe for one to use it and its product when pregnant or breastfeeding, if one has auto-immune infections, it is advisable to refrain from using the same as they tend to make the immune system more active thus increasing the symptoms of autoimmunity (Helmstädter & Staiger, 2014). Before one is scheduled for a surgery, it may be wise to stop taking or using it two weeks before the scheduled time. This is because the central nervous system may be slowed down when anesthesia is introduced. In most cases, the product is normally used by adults. The doses are up to 1000mg daily for three months. It is important to contact a healthcare provider in order to establish the dosage that is best for a specific condition.

Wolff-Parkinson-White (Syndrome)

Wolff-Parkinson-White (W.P.W.) is an increasingly popular cultivated variety of plants grown by farmers all across the country. It is a hearty plant, which thrives in many different environments, and for this reason, it is an extremely popular crop. Wolff-Parkinson-White is a rhododendron hybrid of the hybridum subspecies, named for W. G. Wolff , J. A. Parkinson, and D. J. White, the horticulturalists who discovered it in 1935. Parkinsonia microphylla, commonly known as desert-thyme, white or silver thyme, or W-P-W, is a species of flowering plant in the verbena family (Verbenaceae) that is native to southern California and northern Mexico. It grows in dry washes and canyon bottoms below 2000 ft elevation. Parkinsonia microphylla subsp. polyadenia has been distinguished by some botanists; however, it intergrades with subsp. microphylla. Some references treat this subspecies as a distinct species, P. polyadenia . Parkinsonia microphy (Gupta et al., 2018).

Wolffia is a very old genus and fossils of these plants have been found in Africa, India, China, North America and Australia. These plants contain no chlorophyll and obtain their nutrition from sources other than photosynthesis. Some species are eaten by fish, frogs and ducks which find them. The fruit is edible for humans if cooked, and is a valuable source of vitamins and minerals. The Wolffia is a flowering plant, and commonly known as water crowfoot and Indian butterwort.

The leaves, stems, and roots are used in a wide range of herbal medicines that treat a variety of conditions affecting the head and digestive system. If a loss of appetite, a painful epigastric discomfort or a headache is present, squirting a few drops of the tincture down the patient's throat will generally give good results. The dose is from 2 to 10 drops. A medication that is used for heart conditions in which a patient is unable to contract the heart muscle, or when a patient's heartbeat is too fast. That is what is known as the Wolff-Parkinson-White syndrome. There are no clear instructions concerning the dosage of the plant and its product thus necessary to contact the health worker, the syndrome is used in describing a relatively common heart condition which cause the heart o beat abnormally fast for some periods.

X

Xylia Kerii

Figure 105: Xylia xylocarpa.

Source: Wikipedia

This wonderful plant is grown in the rainforest of Madagascar for distribution all over the world. This is a naturally grown plant, not a chemically dyes item. This is a rare indoor plant found primarily in Madagascar, but it needs the right conditions to thrive. We are providing you with the proper environment for this plant to succeed. Xylia are easy to care for and can bring an inviting freshness to your living space. The leaves will release a sweet aroma when rubbed, making them lovely for any room of your house. It is a beautiful and very unusual houseplant with an exotic and majestic presence. This plant also has some extraordinary features (Griffiths et al., 2002). In China, it is considered a garden plant, with delicate leaves and large scarlet flowers. It is native to China, Japan and Korea and typically grows between 3 and 5 feet tall. It is also known as Lobed leaf, Garden Xylia and Red branches. The perfect plant for placing at the center of a table or desk and brightening up an interior. The *Xylia kerii* is both an essential ingredient for our candles, but also a beautiful addition to any room it inhabits. Its velvety appearance and the wood-like fragrance is subtle yet addicting. It is extremely strong and used throughout the world as a binder for making weapons as well as a source of dye.

The *Xylia kerii*, or brahmi, is a medicinal plant from the mint family. It is used in holistic medicine and Ayurveda to treat wounds, headache pain, depression, chronic fatigue syndrome and arthritis. In traditional Chinese medicine, it treats headaches, inflammation and chronic fatigue. In addition to these uses, it can be used as a food item or a tea drink for its health benefits. It is used by the Chinese for kidney stones. It bears its name after the Greek physician Dioscorides Xylocarpus, an emollient that grows in southeast Anatolia (Fitzgerald et al., 2020). The fresh bark of the tree is ground for food preparation, where it is boiled to relieve stomach pain, ulcer and rheumatoid arthritis. It has many uses for its resin and uses some components of libidinal energy in some woodcutters, especially to those species that are found in Australia. The plant is a supplement which is used to treat diverse conditions including asthma, diabetes, memory loss, Hepatitis C, Alzheimer's disease, high blood pressure, high cholesterol levels and high oxidative stress. It makes a good aperitif and relaxant, and is a favorite for making teas.

Xylopia Aethiopic

Figure 106: Xylopia aethiopica.

Source: SouthWorld

It is a stunning hardy plant that originated in Ethiopia and is easily cultivated in most soil conditions, making it a perfect welcoming gift for anyone. The plant can grow up to a height of 15 feet and has 5-foot long silvery green leaves that resemble bananas. Xylopia's leaves are harvested for their high butyric acid content, a good source of plant-based omega-9 fatty acids. With its refined taste and pleasant smell, this oil can be used in cooking or skin care (Athar & Nasir, 2005). *Xylopia aethiopica* is a species of plant in the citrus family. It is native to Somalia, Tanzania, and Kenya, and is known for its many uses. The Xylopia aethiopica is a plant genus in the family Annonaceae. There are approximately 350 species found throughout tropical regions of Africa, Southeast Asia, Australia, Melanesia and Polynesia, where they grow in a variety of habitats from sea level to 1200 meters. Some species are sources of hardwood timber, many are harvested for their edible fruits, and others are cultivated for ornament. It shows amazing ornamental value to attract birds and butterflies. Its fruit is the source for Xylopia Longhorn Beetle, which is an insect that can pollinate durians. It was also used as a traditional plant medicine to treat skin infections.

Easily grown in the tropics, the medicinal properties of this medicinal plant are used to treat flu and colds. It is easy to grow and is one of the most popular medicinal plants used in traditional Chinese medicine. It is a plant full of benefits. It can be used as astringent, anti-inflammatory, sedative (mental and physical), deodorant, antimicrobial, antipyretic (reduce fever), cooling for external use (rheumatism), antitussive (use in the colds) and relief of itching caused by parasitic fungi or mycosis fungus infections. *Xylopia africana* produces edible fruits which are cooked and eaten as a vegetable dish (Agbogidi & Azagbaekwe, 2013). This plant is used for its astringent properties in the treatment of diarrhea, dysentery, and abdominal pain. It's small clumps of furry leaves are quite meaty and offer a meaty, palm heart-like texture! This tropical farmhouse plant can be grown indoors or outdoors; either way, it will make your home stand out. The fruits are said to be edible and were eaten during famine times by the Akan people in what is now Ghana.

This product should only be consumed in small doses as they can be lethal if consumed in high doses. Taking for instance 5000mg/kg is a very high dose that can be very harmful to the human body.

X heucherella

Habenaria heucherella, also known as the Rose Spectacled Orchid or Lace Orchid, is a terrestrial orchid native to the southeastern United States. The epithet refers to the markings on the labellum, which resemble eyes</ blockquote. It requires little to no maintenance, making it perfect for the novice gardener or busy homeowner. The 10-inch pots allow for increased growing space, enabling you to keep your plant nourished and well-taken care of. This variety has deep green, glossy foliage that looks great in any location, whether indoors or out. From April to June, clusters of white flowers bloom before giving way to bright green glossy foliage that's striking against the rusty brown stems. The X Heucherella plant is a perennial. It grows between 4' and 5' tall. This plant is not damaged by being submerged in water up to its leaves. It can be grown from seed, but may take as long as five years before it blooms. The flowers are a powdery pink, and it blooms in August.

Figure 107: X heucherella.

Source: photograph by Gardenia.net

There is no information regarding the correct doses of X heucherella and thus individuals need to be able to consult their healthcare workers to help them understand the right dosage of the plant.

Y

Yarrow

Figure 108: Both the leaves and flowers of Yarrow have medicinal value.

Source: https://www.gardeningknowhow.com/edible/herbs/yarrow/growing-yarrow.htm

Yarrow is a plant known for its medicinal value. It is also a ground plant and was used traditionally to make medicine. Other names used to refer to yarrow include yarrow essential oil, Achilee, thousand-leaf, Staunchweed,

Soldier's wound wort, Noble yarrow, wound wort, Tausendaugbram, Sourcil de Venus, Sanguinary, Millefolii, Old man's pepper, Roga Mari, Millegoglie, Rajmari, Katzenkart, Millefolium,Milieenrama, Hule Essentielle d'Achile'e, Milefolio, milefolii Floss, common yarrow, Carpenter's weed, Devil's plaything, Devil's nettle, Green arrow, Bloodwort, Civan Percemi, Band Man's plaything and Bauchweh. Various parts of the plants are used for its medicinal value. The leaves of the plant are freshly plucked and chewed to relieve tooth pains. For those suffering from skin problems such as hemorrhoids, a yarrow was applied to the affected areas. It was also used in baths for those suffering from very painful, lower pelvic cramp-like conditions among women (Bussmann & Sharon, 2006). Yarrow was used alongside other herbs in treating stomach problems such as gastrointestinal cramping, intestinal gas and bloating among other problems. Both the leaves and flowers of the yarrow plant are used as food in salads. Yarrow is also used in industries in the making of cosmetic products such as cleansers and shampoos. Other medicinal uses of yarrow include it being used in treating fevers, diarrhea, dysentery, appetite loss, stomach discomfort, common cold, induction of sweating and hay fever. Yarrow is also an essential component used in making mouth wash and is said to help improve cases of gum disease.

Yarrow is a naturally occurring plant and appropriate dosages have not yet been established. However, there some factors that affect the dosage given to patients. They include underlying health conditions, health and age of the patient. The challenge with using yarrow is that scientists have not been able to come up with appropriate dosages. For this reason, patients are advised to bear in mind that yarrow may not be entirely safe. This means that reasonable amounts of dosage should be administered. For products containing yarrow, people should use them according to the prescription. It is also advisable to seek the consult of a physician before consuming such products. Also yarrow is considered safe consumption, possible side effects could be noted when consumed in large amounts. Yarrow products containing thujone are not entirely safe. When consumed orally in medicinal amounts, yarrow is safe for consumption (Eddouks et al., 2017). However, there are certain groups of people who will feel drowsy and an increased urge to urinate when consumed orally. When applied to skin, those with sensitive skin may feel some irritation. Consumption of yarrow among pregnant women may not be safe. Yarrow is known to affect the menstrual cycle. Consumption also increases risk of miscarriages. Lactating mothers are also advised not to use yarrow as there isn't sufficient information to prove that it is safe for them. Another possible side effect of yarrow is decreased blood clotting thereby increasing the possibility of bleeding disorders. It can also cause an allergic reaction for people having sensitivity issues to plants found in the compositae family. It is also advisable that any individual is scheduled to undergo surgery to avoid consumption of yarrow to allow easy management of bleeding.

Some of the yarrow extracts include sunflower oil and olive oil yarrow extracts. These extracts are mostly used in treating skin wounds and dealing with inflammatory issues.

Yerba Mate

Figure 109: Yerba mate is used to make herbal tea.

Source: https://www.coolgarden.me/yerba-mate-plant-growth-and-use-5374/#google_vignette

Yerba mate is a plant whose leaves have medicinal value. The plant is composed of caffeine and other chemicals used in the stimulation of body parts. Other names used to refer to Yerba mate include Green mate, Yerbamate, Chimarrao, Jesuit's, Jesuit's Brazil tea, The' du Bre'sil, Paraguay tea, St. Bartholemew's tea, Ilex paraguariensis, Hervea and Ilex. There are several was Yerba mate was used in treating diseases. The plant is rich in nutrients and antioxidants. This is useful in preventing several diseases. It is also used in boosting mental focus and energy. Yerba mate contains a significant amount of caffeine which helps improve energy as well as mental focus. Yerba mate is also used by athletes to help improve their physical performance. Caffeine found in Yerba mate is said to help muscle contractions while minimizing fatigue (Blancke, 2016). Yerba mate also contains a number of components that can be used to reduce risks of infections caused by fungi, parasites and bacteria. Yerba mate is known to boost metabolism and lower appetite. It is therefore useful in weight loss and getting rid of belly fat. It also contains anti-inflammatory properties that can help boost the immunity system. It is used to lower blood sugar and in the long run reduce the risks of diabetes. The antioxidant properties in Yerba mate are useful in lowering the risks of heart disease.

A number of factors contribute to the dose of Yerba mate. They include underlying diseases, health and age of the patient. Well defined range of dosage has not been established by scientists. Patients should consume a considerable range of dosages. It is also important for them to follow prescriptions indicated on Yerba mate products. Some of the possible side effects of consuming Yerba mate orally include insomnia. Yerba mate contains significant amounts of caffeine linked to insomnia. It also causes increased heart rate, vomiting, nausea, restlessness, upset stomach, vomiting, increased breathing and nervousness. Some of these side effects are brought about by consuming large amount of Yerba mate for a large duration of time. For this reason, it is advisable to take it is small amounts as per the prescription. Consuming large amounts of Yerba mate for extended periods is said to increase the incidence of certain cancers such as mouth, lung, kidney, stomach, cervix, bladder, esophagus and larynx cancer. The incidence of occurrence of these cancers is higher among people who consume alcohol. Due to the caffeine found in Yerba mate, consumption of large amounts causes irregular heartbeats, headache, ringing in the ears, anxiety and agitation. Organic Yerba mate and Yerba mate liquid extracts are some extracts of Yerba mate.

Ylang Ylang

Figure 110: Ylang ylang oil is a common essential oil for Ylang ylang.

Source: https://nayaglow.com/blogs/news/7-things-to-know-about-ylang-ylang

Ylang ylang is a plant whose flowers are used in making medicinal products such as Ylang-ylang oil. Other terms used to refer to Ylang ylang include Huile d'llang-Ilang, Aceite de Ylang Ylang, Cananga odorata forma. Genuina, Canananga odorata, Huile d'Ylang Ylang and Canangium odoratum forma.genuina. *Ylang ylang* is used in making Ylang ylang oil which is used in improving cognitive functions. This includes thinking and memory skills. The oil is used in aroma therapy when a memory test is underway. The oil is also used in getting rid of head lice. The oil is sprayed on the head of the patient and it was highly effective. It is also used in dealing with cases of high blood pressure, promoting relaxation and increasing sexual desire. Application of *Ylang ylang* oil was said to be effective in killing bacteria on skin.

Appropriate dosing of *Ylang ylang* has not been established. There a number of factors that affect dosing of *Ylang ylang* products. They include underlying health conditions, health and age of the affected individual. As it is a natural product, it is advisable that patients take considerable amount of the dosage. In the case of Ylang products, directions on the labels should be strictly adhered to. Professional guidance should also be sought for before using the drug (Dhama et al., 2018). *Ylang ylang* oil among other products are safe for consumption among children and adults. When combined with other herbs, it is safe. However, for individuals with sensitive skin or allergic to plants from the family of plants containing *Ylang ylang*, it causes an allergic reaction. Oral consumption of large amounts of *Ylang ylang* products may cause some side effects. *Ylang ylang* oil is an example of *Ylang ylang* extract.

Yarrow, Mifoil

Figure 111: Mifoil Yarrow is of the Asteraceae family.

Source: https://garden.lovetoknow.com/wiki/Milfoil,_Yarrow

Mifoil Yarrow commonly known as Achillea millefolium is a plant known for its medicinal properties. It was originally grown in northern Asia, America, and Europe. It can also be referred to as the common Yarrow. In some countries, it is considered a common weed and is fed by livestock. Other names for Mifoil Yarrow include *Achillea albida* wild, *Achillea ambigua Pollini*, *Achillea borealis*, *Achillea fusca* Rchb, *Achillea gracilis* Raf, *Achillea lanata* Lam, *Achillea lanulosa* Nutt and *Achillea bicolor* Wender. Mifoil Yarrow is used in drying wool. With regards to traditional medicine, the plant is used in treating ear and toothaches. Leaves of the plant were chewed to help reduce tooth pain while infusions were made to deal with ear pain. In other countries such as California, the plant was used in treating colds. Due to its properties, the plant was used as an analgesic. In some American countries, the plant was used in treating throat infections, abrasions and healing cuts. In other countries, the plant was used to help induce sleep and treat fever and pains. The leaves of the plant were also placed on hot stones and the aroma was inhaled. This helped cure headaches.

As natural medicine, scientists have not been able to develop suitable ranges of dosage. Underlying health conditions, health of the patient and their age determines the type of dosage they are given. Directions written on Mifoil Yarrow products should be followed. Pharmacists should also be consulted before starting a given medication. Mifoil Yarrow is considered safe consumption, possible side effects could be noted when consumed in large amounts. Mifoil Yarrow products containing thujone are not entirely safe. When consumed orally in medicinal amounts, yarrow is safe for consumption. However, there are certain groups of people who will feel drowsy and an increased urge to urinate when consumed orally. When applied to the skin, those with sensitive skin may feel some irritation.

Yellow Dock

Figure 112: The roots of a Yellow dock plant are used for their medicinal value.

Source: https://www.thespruce.com/yellow-dock-plants-2132938

Yellow dock is an herbal plant whose parts are utilized for their medicinal value. Its chemical composition allows it to be used in the medical field. It contains anthraquinones and other chemicals used in treating a different diseases and infections. Other names for Yellow dock include Sour Dock, Curled Dock, Acedera, Chukkah, Field Sorrel, Rumex, Romaza, Sheep Sorrel, Amalvelas, Lengua de Vaca, Broad-Leaved Dock, Narrow Dock, Patience Cre'pue and curly Dock. Yellow dock is used among people who experience swelling and pain in their respiratory tracts and nasal passage. Yellow Dock can be used as a tonic and a laxative. It is used in treating sexually transmitted diseases, arthritis, fungal infections, intestinal infections and bacterial infections (Devkota & Watanabe, 2020). In some areas, Yellow Dock is used as toothpaste. When applied to the skin, Yellow Dock is said to treat skin diseases, rashes, hemorrhoids, dermatitis and stop bleeding. It is also used in treating psoriasis, obstructive jaundice and scurvy among other conditions. It is also used in the treatment of anemia as it is a good source of plant-based iron. The plant can take up significant amounts of iron from the ground enabling its use in treating mild cases of anemia. Its chemical composition includes bio-flavonoids which help improve blood circulation in the body. Yellow Dock is also used in improving immune support. The leaves contain high amounts of vitamin C which ensure our immunity system is healthy. The roots of Yellow dock is used in treating a stressed immune system. An immunity system can be stressed by lymphatic inflammation. The herbs is also used as a skin toner as its chemical properties allow the removal of makeup, perspiration traces and dirt from skin surfaces. It is also used to reduce facial redness and ensure good digestive health. Yellow dock improves the quality of hydrochloric produced in the stomach. Its laxative properties increase digestion allowing frequent emptying of bowels. Yellow dock cleanses and soothes the digestive tract with minimum disturbances. By so doing, the normal pH is maintained and all the useful bacteria in the gut are maintained. It also encourages the release of water into the intestines.

Certain ranges of Yellow Dock dosage have not yet been established by scientists. This is mostly attributed to the lack of sufficient knowledge. Some of the factors affecting the administration of Yellow Dock medications are underlying health conditions, previous diseases, use's age and health. However, as a natural product it is advisable to consume small amount of the drug. Directions listed on products should be followed and a pharmacist consulted before consuming any medication. Side effects vary depending on the mode of usage (Douwes et al., 2008). When consumed orally, some of the possible side effects include low blood levels, diarrhea, stomach cramps, skin irritation and increased urination. When consumed raw, some of the possible side effects include difficulty in breathing, heart problems, vomiting and in extreme cases death. When applied to the skin, Yellow Dock can cause skin irritation. Among pregnant women, Yellow Dock can cause laxative effects that can be passed to the child through the milk. It can also cause an allergic reaction among people who are allergic to ragweed. It irritates the stomach lining and worsen ulcer symptoms.

Yellow Jessamine

Figure 113: Yellow Jessamine plants are known for their yellow flowers.

Source: https://www.istockphoto.com/photos/yellow-jessamine

Yellow Jessamine is a plant whose roots and rhizome are used in the making of medicine. It is bitter. Other than its medicinal value, Yellow Jessamine is used to add flavor to puddings, beverages, baked products, candy and frozen dairy desserts. It is also used as a component in the manufacture of perfumes, lotions and creams. It gives them a characteristic smell. Other names for Yellow Jessamine include woodbine, trumpet flower, false jasmine, Caarolina jasmine, Gelsemin, Evening Trumpet Flower, Gelsemium nitidum, Gelse'mie Toujours Verte, Gelsemio, Gelsemii, Gou Min, Gelsemium Toujours Vert, Jasmin Jaune, Gelsemium sempervirens, Gelsemiumwurzelstock Jessamine, Jasmin de Virginie and Yellow Jasmine. Yellow Jessamine has been used to treat diseases and pain related to the liver such as hepatitis and liver cirrhosis (Do et al., 2019). It is also used to relieve abdominal pains caused by dysentery. The plant is also used in cancer treatment, to heighten sexual desire and as a sedative. Yellow Jessamine is also said to improve mental alertness. This is when the aroma of the plant is inhaled. However, the odor does not affect human concentration. Its roots are said to be vasodilators, analgesic, febrifuge, diaphoretic, mydriatic and hypnotic. The roots are used in treating neuralgia, toothache, insomnia, severe pain, migraine and sciatica.

User's age, health and underlying health conditions are some of the factors that influence the dosage given to a patient. However, appropriate dosing of Yellow Jessamine has not been established due to insufficient information. A small amount of dosage are recommended to avoid the occurrence of any possible long-term or short-term effects. Directions on products should be adhered to. It is also advisable to consult a pharmacist before using the product. Yellow Jessamine is safe to human health when consumed in small amounts. Some individuals may experience some allergic reactions when they consume Yellow Jessamine. It also causes paralysis of the spinal cord, dizziness, double vision, headache, muscle problems, difficulty swallowing, difficulty breathing, seizures, weakness, fever and in severe cases death. Gelsemium is an extract of Yellow Jessamine.

Yerba Mansa

Figure 114: The fruits of *Yerba mansa* are spiky in nature.

Source: https://waterxscapes.com/water-gardens/pond-plants/yerba-mansa-pine-cone-flower-anemopsis-californica/

Yerba mansa is a plant whose rhizome and roots are known for their medicinal properties. Its fruit is usually Reddish-brown in color. The plant has been used in the making of different medicines used in treating a variety of diseases. Other names for *Yerba mansa* include *Yerba manza*, *Anemia californica*, Lizard's tail, Swamp root, *Anemopsis californica*, Queue de Le'zardand Anemopsis de Californie. The plant has been used to treat a variety of diseases among the common cold, gout, cancer, flu, cancer of the uterus, infertility issues more so with regards to the inability to get pregnant, cancer, cough, wound healing, diaper rash, indigestion, pain, vaginal yeast infections, kidney stones, tuberculosis, inflammation of the vagina, menstrual cramps, sexually transmitted diseases, inflammation of the nasal cavity, osteoarthritis and sinuses (Cummings & Olsen, 2011). With regards to cancer, the plant is said to slow down the growth rate of certain cancer cells. It also reduces the amount of mucus produced. Fresh leaves were chewed to make a poultice and put on burns. A bath with leaves of Yerba mansa was said to treat achy feet.

Traditionally Yerba mansa was used extensively to treat diseases. The roots of the plant were chewed and used to prevent infections of the mucus membranes. The roots were used to prepare tea which when consumed was said to clean or purify the blood. Among other communities, the tea was used as wash for wounds and a cure for urinary disorders, asthma and colds. Decocted roots were mashed and used as an anti-septic wash and poultice for swelled areas. Root decoction was also used as a laxative. It was used to relieve an upset stomach. The leaves and roots were used in treating gonorrhea. In some communities, a warm bath using the leaves helped treat fatigue. Decoction of the roots was internally consumed by women to deal with menstrual cramps. Cystitis, painful inflamed joints, sore muscles and chronic lung problems were treated using Yerba mansa. Plant infusion was used to treat stomach ulcers and chest congestion. As an herb, the plant was used to prevent the buildup of uric acid crystals in the kidneys and kidney stones. The leaves were also utilized as a wash for ringworm, blisters, sunburn, poison ivy and insect bites. In Spain, the plant is used in treating skin diseases and blood disorders. It was also used in checking bleeding rate after child birth and treating venereal sores.

Recommended dose range for Yerba mansa has not been established. Dosage is administered depending on a number of factors including underlying health conditions, user's health and user's age. Some underlying health conditions may render a patient unable to consume Yellow mansa. Dosage is administered according to the prescription by the pharmacists. Directions indicated on Yerba mansa products should be followed. Some of the side effects of *Yerba mansa* when consumed orally include drowsiness and sleepiness. When applied on the skin it could cause skin irritation. For people who are allergic to plants of the same family as *Yerba mansa* might experience and allergic reaction. When consumed alongside other drugs it could cause other side effects.

Yohimbe

Figure 115: Yohimbine is an active element in most Yohimbe products.

Source: https://www.lybrate.com/topic/benefits-of-yohimbe-and-its-side-effects

Yohimbe is a tree native to western and central Africa. It is evergreen and its bark is used in the making of medicine. A chemical known as Yohimbine is extracted from the bark of the plant. The chemical has high medicinal properties. It is used alongside other chemical to make drugs. Yohimbine is used in making medicinal supplements in which Yohimbe is the main active ingredient. Other names for Yohimbe include Yohimbehe, 11-hydroxy Yohimbine, Yohimbinum Muriaticum, Corynanthe Johimbe, Corynanthe Yohimbe, Pausinystalia johimbe, Corynanthe yohimbi, Yohimbine HCl, Johimbi, Yohimbehe, Pausinystalia yohimbe and Yohimbehe Cortex. The plant is used to solve sexual problems caused by the consumption of antidepressants. Such problems are referred to as anti-depressant induced sexual dysfunction (Carvalho & Morales, 2010). The bark of a Yohimbe tree contains Yohimbine which is useful in enhancing sexual problems caused by specific antidepressants. Yohimbine in Yohimbe causes an increase in nerve impulses and blood flow to the vagina or penis thereby counteracting effects of anti-depressants. Yohimbine is also used by athletes to improve athletic performance. The Yohimbine found in the bark which helps athletes increase athletic performance. Yohimbine is said to be highly effective in treating anxiety brought about by certain phobias.

The anxiety treatment works well for certain phobias. Studies showed that the plant when used alongside exposure based therapy in treating flying phobias was not effective. However if the combination is used in treating claustrophobia the outcome was much better compared to using Yohimbine alone. Yohimbine is also used in treating dry mouth. When Yohimbine is used, symptoms of dry mouth improved. It is also used in improving depression symptoms. Yohimbine in plants was said to help improve weight loss and erectile dysfunction. Studies show that when people suffering from head rush take Yohimbe, the Yohimbine in it causes an increase in blood pressure. This is as a result of low blood pressure. Yohimbe is quite unique in that scientists have been able to establish a definitive dosage. When consumed by mouth 15-30 mg of Yohimbine is consumed daily to help with issues of sexual performance. The maximum amount of dosage used is 100mg. Higher doses are said to cause side effects and in extreme cases could lead to death.

Consumption of Yohimbe causes side effects ranging from mild to severe. In severe cases, Yohimbe causes heart attacks, seizures, kidney failure, and irregular heartbeats among others. Yohimbe contains Yohimbine which is the active ingredient in most medicines. When used for a short period of time under the supervision of a pharmacist or a medical practitioner. Without proper supervision, consumption of Yohimbe could result in severe side effects (Bonini et al., 2018). Typical doses of Yohimbe can cause vomiting, stomach upset, drooling, irritability, excitation, frequent urination, tremor, agitation, anxiety, headache, nausea, sleep problems, sinus pain, bloating, high blood pressure, rash, racing heartbeat, stomach pains and dizziness. Sever problems brought about by consumption of high amounts of the drug include heart problems, low blood pressure, paralysis, difficulty breathing and in extreme cases death. It was noted that the day after consuming Yohimbine patients reported case of allergic reactions alongside kidney failure, chills, scaly skin, itchy skin, and listlessness. Some experienced lupus.

Yucca

Figure 116: The roots of Yucca have strong medicinal value.

Source: https://www.gardendesign.com/succulents/yucca.html

Yucca is a plant have belong to the Yucca genus. There about 40 species of plant under the Yucca genus. The plant does not bear any flower. Other names for Yucca include Yucca whipplei, Adam's Needle, Yucca schidigera, Yucca arborescens, Yucca glauca, Yucca mohavensis, Yucca brevifolia, Yucca filamentosa, Aloe Yucca, Bear Grass, Yucca angustifolia, Spanish Bayonet, Soapweed, Yucca, Dagger plant, Mojave Yucca and Arbre de Josue'. Yucca is used in the treating intestinal inflammation, migraine headaches, high cholesterol, gallbladder disorders, liver diseases, osteoarthritis, diabetes, high blood pressure and stomach disorders. Skin disorders such as dandruff, baldness and skin diseases are treated by applying yucca on the skin. Yucca is also used in the treatment of joint pain, sprains and sores (Bhattacharyya, 2015). When treating high cholesterol and high blood pressure chemicals found in Yucca are actively involved in controlling the problem. It is also used in treating symptoms of arthritis such as stiffness, pain and swelling. Other than treating with issues of high blood pressure, Yucca is used in treating heart disease. The plant helps by minimizing oxidative stress placed on the cardiovascular system. The oxidative stress is caused by a lack of balance between antioxidants and free radicals. Normal metabolic processes generate the free radicals and are known to damage DNA, proteins, lipids and cell linings. By damaging cells at the genetic level, they cause heart disease and cancer among other diseases. By minimizing oxidative stress, Yucca helps reduces risk of heart disease.

Scientists are have not established the appropriate range of dosage for Yucca. An appropriate dosage is administered according to number of factors including underlying health conditions, user's age and their health. Knowledge that Yucca is a natural product is useful in helping consumers known what they are dealing with. It is crucial for pharmacists to be consulted before the medication is started. Directions on product labels should be followed. It is also important for consumers to know the type of Yucca dietary supplements they are ingesting. This is because most of the supplements are not regulated, tested or subjected to routine safety by government board. It is advisable for one to identify a brand known to regulate their products and have a well-established market presence. It is also good to check whether the supplements are intended for animals or for humans. Animal supplements may contain larger or lower amounts of Yucca.

Short-term consumption of Yucca is safe for human beings. Some individuals may experience stomach upset. There is insufficient knowledge on possible side effects of long-term oral and dermal consumption of Yucca. When applied to the skin, some of the possible side effects include skin rash and allergic reactions.

In some cases, consumption of Yucca causes breathing difficulties. It is advised that lactating and pregnant women avoid the consumption of Yucca though there is insufficient information to prove that it is harmful.

Z

Zea Mays

Figure 117: Maize is a good source of carbohydrates.

Source: https://www.alamy.com/maize-corn-zea-mays-botany-book-1850-image332703921.html

Zea mays is a plant easily identified by its long shiny fibers. The fibers are usually at the top of an ear of corn.

The term *Zea mays* is the scientific name for maize plants. It is under the family Poaceae. There are about 10,000 species of crops categorized under this family. In some parts of the world, it is commonly called corn. The plant is characterized by a leafy stalk whose main role is the production of pollen grains. There are different types of Zea mays namely sweet corn, flour corn, popcorn, pod corn, flint corn and dent corn (Braun, 2005). Other than being a dietary staple, *Zea mays* have medicinal value. Parts of the plant are used to make medicine most common part being the corn silk. Corn silk is the long shiny fibers at the top of ear of corn. Other names used for Zea mays include Corn silk, Maize silk, Indian corn, Zea, Barbe de Mais, Stigma Maydis, Cheveux de Mais, Soie de Mais, Maidis Stigma, Yu Mi Xiu and Purple corn.

Zea mays is used in stimulating urination. It is considered a diuretic as it possesses the ability to increase urination. This is useful in the removal of excess potassium and toxins from the body. By stimulating the release of wastes, it reduces the risk of an inflamed urinary system. *Zea mays* is also used to support kidney health by reducing chances of an individual suffering kidney stones. It can get rid of kidney stones. The chemical composition of *Zea mays* contributes to its use in treating issues of blood sugar levels. *Zea mays* is said to influence blood sugar levels in the body and ensure that they are balanced. It is also used in treating high blood cholesterol. Some of the elements in Zea mays lower blood cholesterol and in the long run protect the body from heart-related diseases (Maruca et al., 2019). It is also used in treating bed-wetting issues, fatigue, diabetes, inflamed prostate and congestive heart failure. It is also used in ensuring good health among individuals. *Zea mays* contain a significant amount of dietary fiber, carbohydrates, manganese, copper, zinc, potassium, phosphorus, magnesium and plant-based proteins. Sufficient amounts of these elements help in developing good sleeping patterns. It is also a good source of vitamin B3 (which is very useful in ensuring proper functioning of nerves, skin and the digestive system), B5 (needed in ensuring the metabolism of both proteins and fat, choline, vitamin B9, thiamine and Vitamin B6.

Doses of *Zea mays* can be administered in form of capsules, oil, ground, decoction, infusion, cooked, herbal supplements or tincture. There a number of factors that influence the dose of *Zea mays* given to individuals. They include underlying health conditions, user health and user age. The lack of sufficient knowledge contributes to the undefined dosage of *Zea mays*. It is important that users seek the advice of a professional before starting the dosage. They should also follow directions indicated on Zea mays products. In as much as there is no information on appropriate dosage ranges, *Zea mays* is used alongside other medications to treat a variety of diseases. For instance with the treatment of diabetes, *Zea mays* can be used alongside anti-diabetes drugs. However, much caution should be taken as combination of both medications could extremely lower blood sugar levels (Martínez & Luján, 2011). In the treatment of high blood pressure, *Zea mays* is consumed alongside anti-hypertensive drugs such as losartan and captopril. The combination of the drugs lower blood pressure. The same applies when dealing with cases of inflammation. *Zea mays* is used alongside Coumadin. When Coumadin is consumed in large amount it makes the blood clotting process really slow. Zea mays of the other hand helps improve clotting of blood. Therefore when *Zea mays* is consumed alongside Coumadin, it counteracts the effects of Coudamin so that the clotting process is not very slow. *Zea mays* is also used alongside water pills. Water pills are considered Diuretic drugs. Both *Zea mays* and water pills almost play the same role. Both the drugs cause the body to get rid of both water and potassium. Taking the combination will cause a significant decrease in the amounts of potassium in the body. While taking both *Zea mays* and other drugs, a pharmacist must be consulted.

Zea mays is generally safe for consumption. However, if individuals are allergic to plants under the family of plants similar to *Zea mays* it will cause an allergic reaction. Other possible side effects include itching and skin rashes as *Zea mays* cause a decline in potassium in the body. Other possible side effects include heart palpitations, muscle cramping, weakness, nausea and numbness. Pregnant women are advised not to consume medicinal doses of *Zea mays* as it could lead to miscarriages. The same case applies to lactating mothers. Medicinal doses could disrupt normal milk production. Supplements of *Zea mays* are not advisable for individuals suffering from corn allergies, low potassium, blood pressure issues and diabetes. For patients suffering from diabetes, precaution is necessary as significantly large amounts of *Zea mays* could lower blood sugar levels. This could eventually disrupt the control in blood sugar. The same case applies to people suffering from blood pressure issues. Extremely large amounts of *Zea mays* could modify the management of the condition. Zea mays silk extract is an example of *Zea mays* extract.

Zingiber Officinale

Figure 118: Zingiber officinale is known to treat nausea and pain.

Source: https://www.medicalnewstoday.com › ... Web results Ginger water: Benefits, risks, and how to make it at home

Zingiber officinale commonly known as Ginger is a plant whose medicinal value is well known around the world. It was originally grown in Asia. The root of the plant was commonly used in the making of spices. It chemical composition allowed it to be used in the treatment of a variety of diseases. Other names for *Zingiber officinale are ginger, Zinziber officinalis*, Shunthi, Shoga, African ginger, *Zingiber officinale*, Amomum Zingiber, Zinzeberis, Ardaka, *Zingiberis Siccatum Rhizoma*, Black ginger, Zingiberis Rhizoma, Cochin Ginger, Vishvabheshaja, Gan Jiang, Nagara, Race ginger, Ginger root, Ginger essential oil, Gingembre and Kanshokyo.

Zingiber officinale is used in treating swelling and nausea. Chemicals found in ginger can control the nervous system preventing nausea. It is mostly used by patients suffering from HIV/AIDS to get rid of nausea and vomiting caused by the medication. It is also used in treating menstrual cramps. When consumed in the initial days of the menstrual cycle, *Zingiber officinale* helps reduce the pain of menstrual periods (Mati & de Boer, 2011). Some say that *Zingiber officinale* works efficiently like other menstrual drugs such as ibuprofen or Novafen. In some cases, Zingiber officinale is taken alongside these medications. *Zingiber officinale* is used among pregnant women to treat morning sickness. It gets rid of the vomiting and nausea. It has also been used in treating Osteoarthritis. People suffering from osteoarthritis can consume *Zingiber officinale* to help minimize the pain. In some cases, ginger gel when applied to the knee has proven helpful. *Zingiber officinale* is also used as a cure for migraines.

Zingiber officinale is unique when it comes to dosage as it is quite different from other plants. Appropriate ranges of dosage have been established. There are various forms of consuming *Zingiber officinale* as medicine. It can be ingested as a liquid extract, capsule, Syrup and tea. For adults, oral dosage is between 0.5-3 grams for 12 weeks. Aromatherapy essential oils, ointments and topical gels are *Zingiber officinale* products used to treat diseases.

When taking *Zingiber officinale* to treat vomiting and nausea among HIV/AIDS patients, 0.5-3 mg of *Zingiber officinale* is taken 30 minutes before their treatment. For menstrual cramps, 0.5-3 mg of Zingiber officinale is taken for 3-4 days of the cycle. *Zingiber officinale* is used alongside other drugs in treating a variety of diseases. It is used alongside anti-platelet drugs, Coumadin, Phenprocoumon and Calcium channel blockers among others. Some of the possible side effects of ginger when consumed orally include stomach discomfort, burping, diarrhea and heartburn. Higher risks could be incurred if the patient has a dose larger than 5 mg. When applied on the skin, some of the side effects could be skin irritation. Long-term usage could cause other severe effects (Li et al., 2015). For pregnant women, they are advised on to consume *Zinginber officinale* when getting close to the delivery date. This is because it increases the risk of bleeding. It is useful in the initial stages of pregnancy. Zingiber officinal also increases the risk of bleeding disorders and can worsen heart conditions. An indivdual set shouldn't undergo surgery to consume Zingiber officinale as it could slow down the blood clotting process.

Zinc

Figure 119: Zinc is a crucial element used in the body's development.

Source: https://www.greenwaybiotech.com/blogs/gardening-articles/whats-the-function-of-zinc-zn-in-plants

Zinc is a nutrient found in plants and soil. It is extracted from plants and used in making a variety of medicines. It also used in medical research. Zinc is mostly useful in a growing child as it helps them develop their immunity system. It is also used to support a variety body functions more so in the making of DNA and proteins. Zinc is richly found in certain plants among them whole-grain cereals, nuts and beans. Other names for zinc include Zinc Monomethionine, Zinc Methionine, Zinc Methionate, Zinc Gluconate, Zinc Difumarate Hydrate, Gluconate de Zinc, Acexamate de Zinc, Zinc Chloride, Ace'tate de Zinc, Zinc Citrate, Citrate de Zinc, Zinc oxide, Zn, Zinc Aspartate, Zinc Acetate, Zinc Sulphate, Zinc Oxide, Zincum Metallicum and atomic number 30 (López-Malo et al., 2020).

Zinc is mostly used among people suffering from asthma, lung infections, malaria and the common cold. It is also used to treat age-related problems such as Alzheimer's and muscle degeneration. Zinc is used to lower the risk of pneumonia among other infections by ensuring the immune system is properly functioning. Infants suffering from diarrhea are given zinc to shorten the diarrhea bouts. Zinc plays an essential role in ensuring that our skin is healthy. For this reason, zinc is used to facilitate wound healing. It is also recommended for individuals suffering from ulcers as it helps treat the wounds. Zinc contains antioxidant properties useful in reducing stress. Studies show that there is a link between chronic diseases and chronic stress. Therefore by dealing with oxidative stress, the risk of suffering from metabolic syndrome, diabetes and high blood pressure are low. Zinc is said to slow down the process of age-related macular degeneration. It protects the retina from cell damage and vision loss. Zinc is used in treating sexual health issues more so in men. Zinc is useful in ensuring proper sexual development at the appropriate time (Kujawska et al., 2017). However, the amount of zinc consumed should be controlled to avoid damage to the sperms. Studies show that zinc could be used in the treatment of diaper dermatitis, acne vulgaris, atopic dermatitis and hidradentitis suppurativa. It is also used to treat neurological symptoms such as peripheral neuropathy and headache. Most of these symptoms are caused by zinc deficiency in the body. Alongside neurology, zinc is also useful in boosting cognitive behavior. It increases memory and thinking capacity.

Zinc dosages vary depending on the age of the patient and their gender. For children of the age 0-6 months, dosage may vary from 0.5 to 2 mg. This applies to both male and female infants. Children of the age 7-12 months dosage is up to 3 mg for both the genders. The same applies to children of ages 1-3 years. For children of ages 4-8 years, treatment ranges from 0.5-5 mg for both genders while those of ages 9-13 can receive dosages ranging from 0.5-8 mg for both genders. For those of ages 14-18 years males receive a dosage of up to 11mg of zinc while females receive dosages of up to 9 mg of zinc. Individuals above 19 years receive a dosage of up to 11 mg (for males) while females receive 8mg of Zinc. Some of the possible side effects of consuming Zinc include an upset stomach, nausea and bad taste (Kunwar et al., 2016). When ingested orally, some of the side effects include kidney damage, stomach damage, diarrhea, vomiting and metallic taste. For an adult, the maximum dosage of zinc they can consume is about 40 mg. when consumed for longer durations of time, zinc could build up to toxic levels. For this reason, routine zinc supplements are not recommended. When the ingested dosage is above 40 mg, the body will react by reducing the amount of copper being absorbed thereby causing anemia. Exceedingly high amounts of zinc also causes fatigue, stomach pains, coughing and fever. Consumption of more than 100 mg of zinc supplements increase the risk of suffering from prostate cancer. When 100 mg of zinc supplements are consumed alongside multivitamins, the risk of dying from prostate cancer increases. Consumption of about 450 mg of zinc could lead to iron issues with the blood while single doses of 10-30 mg of zinc could cause fatalities. When applied on the skin, some of the possible side effects include tingling, itching, stinging and burning sensations. When inhaled, the possible side effect is loss of smell. In some cases, the loss of smell is permanent.

Zero Moisture Basis

Moisture is an essential element in the growing of plants more so medicinal plants as it contributes to the plant composition. For this reason, farmers heavily rely on information from meteorologists to identify the most appropriate time to plant some of the medicinal plants. There are stations across different states involved in acquiring data on soil moisture. In the stations, researchers use a soil sample and using several parameters can establish the amount of moisture in soil. Studies are done to take into consideration some of the biases more so evening and morning bias errors and factor in how temperature could affect moisture levels in the soil. For plants such as wheat, uncontrolled soil moisture could destroy the crops. Wheat possesses medicinal value (Kovach et al., 2009). For this reason, researchers have been able to develop a mechanism in which wheat can be grown under zero moisture basis. The zero moisture basis is also used in the marketing of wheat and its products. Moisture is said to affect the amount of protein found in plants. By growing wheat on a zero moisture basis, they become more marketable. This has been noted in most American states. The zero moisture basis could also be useful in ensuring that the concentration of certain elements in plants are not affected. In other words, they help improve the levels of these nutrients and certain chemicals in plants. This makes the medicinal components readily available and can be used to make a variety of medicinal products and supplements. The use of zero moisture basis in growing wheat has been used as a reference method for other grains and plants. This method of planting has been embraced in other countries and is becoming more popular across different parts of the world.

Zedoary

Figure 120: Zedoary is commonly grown in Thailand.

Source: https://www.dreamstime.com/royalty-free-stock-photos-zedoary-luya-luya-shan-curcuma-zedoaria-berg-roscoe-plant-herb-thailand-health-image30468898

Zedoary is a plant whose rhizome is utilized traditionally in the making of medicine. The plant was prepared by washing the rhizome with plenty of water. This was done to get rid of most of the water-soluble nutrients and proteins. Washing the rhizome was also useful in removing the poison. Other names for Zedoary are ZTO, Zedoary oil, Cetoal, Zitwerwirtzelstock, Amomum zedoaria, Zeodary turmeric, Zedoarie Rhizoma, Gingembre Ba'tard, Kua, Cedoaria, Sati, white turmeric, Temu putih, E-Zhu, Indian arrowroot, Zedoaria, Zedoary oil, Zitwer, Temu Kuning, Shati, Round Zedoary, Kachur, Curcuma Zedoaria, Rhizome de Ze'doaire and Huile de Ze'doaire. Zedoary is used in treating indigestion, coughs, fatigue, stomach pain and cancer. It is also used to relieve stress and treat anxiety. It is a good mosquito repellant and can be used to improve appetite. It is also effective in reducing inflammation more so swelling of the liver (Karmakar et al., 2010). Early research shows that Zedoary is used in treating high cholesterol. This was done by drinking tea made using Zedoary roots. The tea helped lower cholesterol. It was also said to increase high-density lipoprotein cholesterol among men. Traditionally Zedoary was used in treatment due to its hepatoprotective, anti-inflammatory, demulcent, expectorant, antiulcer, rubefacient, antimicrobial stomachic and carminative actions. For this reason it was used therapeutically in relieving fever and hiccups, helping wound healing, removing worms and treating goiter, abdominal lumps and leprosy.

Appropriate dose ranges for zedoary have not been established. The user's health, age and underlying diseases influence the type of dosage given to them. Though it is a natural drug, it is important for caution to be taken. Pharmacist and physician should be consulted before starting treatment. Directions written on Zedoary products should be followed. Before using Zedoary alongside other medications to treat particular diseases, a physician should first be consulted.

Zeodary is a natural plant and is considered safe for consumption when taken orally or when applied to the skin. For this reason, there are no reports of side effects. However, individuals allergic to Zedoary may experience allergic reactions or skin itchiness. It is recommended for pregnant women to avoid Zedoary during the pregnancy period as it could lead to a miscarriage. Lactating mothers are also advised to avoid it as there isn't sufficient information to know the effects of the drug on the infant. Women who experience heavy menstrual periods are advised to avoid Zedoary.

BIBLIOGRAPHY

1. Asnake, S., Teklehaymanot, T., Hymete, A., Erko, B., & Giday, M. (2016). Survey of medicinal plants used to treat malaria by Sidama People of Boricha District, Sidama Zone, South Region of Ethiopia. *Evidence-Based Complementary and Alternative Medicine, 2016.*
2. Athar, M., & Nasir, S. M. (2005). Taxonomic perspective of plant species yielding vegetable oils used in cosmetics and skin care products. *African journal of biotechnology*, *4*(1), 36-44.
3. Agbogidi, O. M., & Azagbaekwe, O. P. (2013). Health and nutritional benefits of nutmeg (Mystica fragrans houtt.). *Scientia Agriculturae*, *1*(2), 40-44.
4. Bussmann, R. W., & Sharon, D. (2006). Traditional medicinal plant use in Northern Peru: tracking two thousand years of healing culture. *Journal of ethnobiology and ethnomedicine*, *2*(1), 1-18.
5. Blancke, R. (2016). *Tropical fruits and other edible plants of the world: An illustrated guide*. Cornell University Press.
6. Bhattacharyya, B. (2015). *Golden Greens: The Amazing World of Plants*. The Energy and Resources Institute (TERI).
7. Braun, L. (2005). Olive-leaf Extract Olea europaea. *Journal of Complementary Medicine: CM, The*, *4*(3), 69-74.
8. Bonini, S. A., Premoli, M., Tambaro, S., Kumar, A., Maccarinelli, G., Memo, M., & Mastinu, A. (2018). Cannabis sativa: A comprehensive ethnopharmacological review of a medicinal plant with a long history. *Journal of ethnopharmacology*, *227*, 300-315.
9. Carvalho, A. M., & Morales, R. (2010). Persistence of wild food and wild medicinal plant knowledge in a northeastern region of

Portugal. *Ethnobotany in the New Europe: People, health and wild plant resources*, 147-171.

10. Cummings, A. J., & Olsen, M. (2011). Mechanism of action of stinging nettles. *Wilderness & environmental medicine*, *22*(2), 136-139.
11. Do, H. D. K., Jung, J., Hyun, J., Yoon, S. J., Lim, C., Park, K., & Kim, J. H. (2019). The newly developed single nucleotide polymorphism (SNP) markers for a potentially medicinal plant, Crepidiastrum denticulatum (Asteraceae), inferred from complete chloroplast genome data. *Molecular biology reports*, *46*(3), 3287-3297.
12. Douwes, E., Crouch, N. R., Edwards, T. J., & Mulholland, D. A. (2008). Regression analyses of southern African ethnomedicinal plants: informing the targeted selection of bioprospecting and pharmacological screening subjects. *Journal of Ethnopharmacology*, *119*(3), 356-364.
13. Devkota, H. P., & Watanabe, M. (2020). Role of medicinal plant gardens in pharmaceutical science education and research: An overview of medicinal plant garden at Kumamoto University, Japan. *Japan. Journal of Asian Association of Schools of Pharmacy*, *9*, 44-52.
14. Dhama, K., Karthik, K., Khandia, R., Munjal, A., Tiwari, R., Rana, R., ... & Joshi, S. K. (2018). Medicinal and therapeutic potential of herbs and plant metabolites/extracts countering viral pathogens-current knowledge and future prospects. *Current drug metabolism*, *19*(3), 236-263.
15. Eddouks, M., Ajebli, M., & Hebi, M. (2017). Ethnopharmacological survey of medicinal plants used in Daraa-Tafilalet region (Province of Errachidia), Morocco. *Journal of ethnopharmacology*, *198*, 516-530.
16. Fitzgerald, M., Heinrich, M., & Booker, A. (2020). Medicinal plant analysis: A historical and regional discussion of emergent complex techniques. *Frontiers in pharmacology*, *10*, 1480.
17. Griffiths, G., Trueman, L., Crowther, T., Thomas, B., & Smith, B. (2002). Onions—a global benefit to health. *Phytotherapy research*, *16*(7), 603-615.
18. Gupta, B. M., Gupta, R., Agarwal, A., & Goel, S. (2018). Ocimum sanctum (Medicinal plant) research: A scientometric assessment of global publications output during 2008–17. *International Journal of Information Dissemination and Technology*, *8*(2), 67-73.
19. Helmstädter, A., & Staiger, C. (2014). Traditional use of medicinal agents: a valid source of evidence. *Drug Discovery Today*, *19*(1), 4-7.

20. Jima, T. T., & Megersa, M. (2018). Ethnobotanical study of medicinal plants used to treat human diseases in Berbere District, Bale Zone of Oromia Regional State, South East Ethiopia. *Evidence-Based Complementary and Alternative Medicine, 2018.*

21. Jiang, D., Jin, X. L., Sun, J. H., Yuan, Q. J., Chiang, Y. C., & Zhang, Z. Y. (2019). Phylogeography of Medicinal Plant. In *Molecular Pharmacognosy* (pp. 55-102). Springer, Singapore.

22. Kassam, K. A., Karamkhudoeva, M., Ruelle, M., & Baumflek, M. (2010). Medicinal plant use and health sovereignty: findings from the Tajik and Afghan Pamirs. *Human ecology, 38*(6), 817-829.

23. Kayani, S., Ahmad, M., Sultana, S., Shinwari, Z. K., Zafar, M., Yaseen, G., ... & Bibi, T. (2015). Ethnobotany of medicinal plants among the communities of Alpine and Sub-alpine regions of Pakistan. *Journal of ethnopharmacology, 164*, 186-202.

24. Karmakar,A.,Karmakar,S.,&Mukherjee,S.(2010).Propertiesofvarious plants and animals feedstocks for biodiesel production. *Bioresource technology, 101*(19), 7201-7210.

25. Kovach, M. J., Calingacion, M. N., Fitzgerald, M. A., & McCouch, S. R. (2009). The origin and evolution of fragrance in rice (Oryza sativa L.). *Proceedings of the National Academy of Sciences, 106*(34), 14444-14449.

26. Kunwar, R. M., Baral, K., Paudel, P., Acharya, R. P., Thapa-Magar, K. B., Cameron, M., & Bussmann, R. W. (2016). Land-use and socioeconomic change, medicinal plant selection and biodiversity resilience in far Western Nepal. *PLoS One, 11*(12), e0167812.

27. Kujawska, M., Hilgert, N. I., Keller, H. A., & Gil, G. (2017). Medicinal plant diversity and inter-cultural interactions between indigenous guarani, criollos and polish migrants in the subtropics of Argentina. *PlOS one, 12*(1), e0169373.

28. López-Malo, A., Alzamora, S. M., Paris, M. J., Lastra-Vargas, L., Coronel, M. B., Gómez, P. L., & Palou, E. (2020). Naturally occurring compounds–Plant sources. *Antimicrobials in food*, 527-594.

29. Li, L., Liu, Y., Wang, X., Fang, J., Wang, Q., Zhang, B., ... & Terwei, A. (2015). Different effects of regional species pool on plant diversity between forest and grassland biomes in arid northwest China. *Plos one, 10*(7), e0131982.

30. Mati, E., & de Boer, H. (2011). Ethnobotany and trade of medicinal plants in the Qaysari Market, Kurdish Autonomous Region, Iraq. *Journal of Ethnopharmacology*, *133*(2), 490-510.
31. Martínez, G. J., & Luján, M. C. (2011). Medicinal plants used for traditional veterinary in the Sierras de Córdoba (Argentina): an ethnobotanical comparison with human medicinal uses. *Journal of Ethnobiology and Ethnomedicine*, *7*(1), 1-19.
32. Maruca, G., Spampinato, G., Turiano, D., Laghetti, G., & Musarella, C. M. (2019). Ethnobotanical notes about medicinal and useful plants of the Reventino Massif tradition (Calabria region, Southern Italy). *Genetic Resources and Crop Evolution*, *66*(5), 1027-1040.
33. Muchugi, A., Lengkeek, A. G., Kadu, C. A. C., Muluvi, G. M., Njagi, E. N., & Dawson, I. K. (2006). Genetic variation in the threatened medicinal tree Prunus africana in Cameroon and Kenya: implications for current management and evolutionary history. *South African Journal of Botany*, *72*(4), 498-506.
34. Menendez-Baceta, G., Aceituno-Mata, L., Reyes-García, V., Tardío, J., Salpeteur, M., & Pardo-de-Santayana, M. (2015). The importance of cultural factors in the distribution of medicinal plant knowledge: a case study in four Basque regions. *Journal of Ethnopharmacology*, *161*, 116-127.
35. Miara, M. D., Bendif, H., Hammou, M. A., & Teixidor-Toneu, I. (2018). Ethnobotanical survey of medicinal plants used by nomadic peoples in the Algerian steppe. *Journal of ethnopharmacology*, *219*, 248-256.
36. Nadda, R. K., Ali, A., Goyal, R. C., Khosla, P. K., & Goyal, R. (2020). Aucklandia costus (syn. Saussurea costus): Ethnopharmacology of an endangered medicinal plant of the Himalayan region. *Journal of Ethnopharmacology*, 113199.
37. Ningthoujam, S. S., Talukdar, A. D., Potsangbam, K. S., & Choudhury, M. D. (2012). Challenges in developing medicinal plant databases for sharing ethnopharmacological knowledge. *Journal of ethnopharmacology*, *141*(1), 9-32.
38. Peng, L. P., Cai, C. F., Zhong, Y., Xu, X. X., Xian, H. L., Cheng, F. Y., & Mao, J. F. (2017). Genetic analyses reveal independent domestication origins of the emerging oil crop Paeonia ostii, a tree peony with a long-term cultivation history. *Scientific reports*, *7*(1), 1-12.
39. Rawat, G., Bameta, A., & Gaur, A. K. (2020). Urtica dioica (Stinging Nettle): A Pharmacologically Important Medicinal Plant Species

of Himalayan Region. *International Research Journal of Pure and Applied Chemistry*, 49-58.

40. Reinaldo, R., Albuquerque, U., & Medeiros, P. (2020). Taxonomic affiliation influences the selection of medicinal plants among people from semi-arid and humid regions—A proposition for the evaluation of utilitarian equivalence in Northeast Brazil. *PeerJ*, *8*, e9664.
41. Saslis-Lagoudakis, C. H., Williamson, E. M., Savolainen, V., & Hawkins, J. A. (2011). Cross-cultural comparison of three medicinal floras and implications for bioprospecting strategies. *Journal of Ethnopharmacology*, *135*(2), 476-487.
42. Schultz, F., Anywar, G., Wack, B., Quave, C. L., & Garbe, L. A. (2020). Ethnobotanical study of selected medicinal plants traditionally used in the rural Greater Mpigi region of Uganda. *Journal of ethnopharmacology*, *256*, 112742.
43. Salim, M. A., Ranjitkar, S., Hart, R., Khan, T., Ali, S., Kiran, C., ... & Xu, J. (2019). Regional trade of medicinal plants has facilitated the retention of traditional knowledge: case study in Gilgit-Baltistan Pakistan. *Journal of ethnobiology and ethnomedicine*, *15*(1), 1-33.
44. Sepúlveda-Cuellar, L., Duque-Ortiz, A., Yáñez-Espinosa, L., Calzada, F., Bautista, E., Pastor-Palacios, G., ... & Douterlungne, D. (2021). Phylogenetic and Chemical Analyses of the Medicinal Plant Salvia circinnata: an Approach to Understand Metabolomics Differences. *Revista Brasileira de Farmacognosia*, 1-13.
45. Silva, N. C. C., & Fernandes Júnior, A. J. J. O. V. A. (2010). Biological properties of medicinal plants: a review of their antimicrobial activity. *Journal of venomous Animals and Toxins including tropical diseases*, *16*, 402-413.
46. Teixidor-Toneu, I., Jordan, F. M., & Hawkins, J. A. (2018). Comparative phylogenetic methods and the cultural evolution of medicinal plant use. *Nature Plants*, *4*(10), 754-761.
47. Tuttolomondo, T., Licata, M., Leto, C., Savo, V., Bonsangue, G., Gargano, M. L., ... & La Bella, S. (2014). Ethnobotanical investigation on wild medicinal plants in the Monti Sicani Regional Park (Sicily, Italy). *Journal of Ethnopharmacology*, *153*(3), 568-586.
48. Nazim, M., Qurat-Ul-Ain Sadiq, A. N., Anjum, S., Ali, M., & Maryam, H. (2020). Mentha arvensis, a medicinal and aromatic plant, has high nutritional value and several-uses: A review. *Buletin Agroteknologi*, *1*(2), 37-49.

49. Saslis-Lagoudakis, C. H., Hawkins, J. A., Greenhill, S. J., Pendry, C. A., Watson, M. F., Tuladhar-Douglas, W., ... & Savolainen, V. (2014). The evolution of traditional knowledge: environment shapes medicinal plant use in Nepal. *Proceedings of the Royal Society B: Biological Sciences*, *281*(1780), 20132768.
50. Seth, M. K. (2003). Trees and their economic importance. *The Botanical Review*, *69*(4), 321-376.
51. Stanisavljević, I., Lakićević, S., Veličković, D., Lazić, M., & Veljković, V. (2007). The extraction of oil from tobacco (Nicotiana tabacum L.) seeds. *Chemical Industry and Chemical Engineering Quarterly*, *13*(1), 41-50.
52. Thakur, G. (2011). *Mimosa Pudica* (Doctoral dissertation, University of Florida).
53. Umadevi, M., Pushpa, R., Sampathkumar, K. P., & Bhowmik, D. (2012). Rice-traditional medicinal plant in India. *Journal of pharmacognosy and phytochemistry*, *1*(1), 6-12.
54. Van Jaarsveld, E. (2013). *Waterwise gardening in South Africa and Namibia*. Penguin Random House South Africa.
55. Yi, Y. J., Cheng, X., Yang, Z. F., & Zhang, S. H. (2016). Maxent modeling for predicting the potential distribution of endangered medicinal plant (H. riparia Lour) in Yunnan, China. *Ecological Engineering*, *92*, 260-269.
56. Zhou, J., Wang, W., Liu, M., & Liu, Z. (2014). Molecular authentication of the traditional medicinal plant Peucedanum praeruptorum and its substitutes and adulterants by DNA-barcoding technique. *Pharmacognosy Magazine*, *10*(40), 385.

INDEX

A

Abelmoschus esculentus 1, 2
Abrus precatorius 3, 4
Abscess root 19
Acacia nilotica 22
acetylcholinesterase 25
Acorus calamus 5
Acyl-CoA 24
Adam's needle 20
Adenosine monophosphate (AMP) 26
Aegle marmelos 35
Allium sativum 12, 13
Aloe vera 13, 14
Alpha-linolenic acid (ALA) 25
Andrographis paniculata 14
Andropogon annulatus 16
Annona muricata 17
Antipyretics 20
Arachis hypogaea 18
Aryl hydrocarbon hydroxylase (AHH) 25
Asthma 223, 224
Astringents 20
Ayurvedic medicine 272, 273, 274

B

bamboo 28
Bamboo shoots 28
Baneberry 36
barks 30
Barleria lupulina 29
Bixa orellana 31, 32
Black Catechu 7
Blumea balsamifera 32
Breynia patens 34
Brown mustard 32

C

Calendula 40
Camellia sinensis 40, 41
Camphor tree 47
Cannabis sativa 42
Capparis tomentosa 23
Catechu 38, 39
Chickpeas 50, 51
chili pepper 49
Clove 71, 72
coconut tree 46
Colforsine 85
Combretum indicum 222

congestive heart failure or CHF 84
Cough 223

D

Daikon 52, 53
Dandelion coffee 54
Decoctions 54, 60
deglycyrrhizinated licorice (DGL) 63
Desmodium triflorum 59, 60
Devil's claw root 56
DHT (dihydrotestosterone) 64
Dill Dong Quai 57
Dioscorea alata 60
Dolichos catjang 61

E

Ecklonia cava 140, 141, 142
Eicosapentaenoic acid 75
Electuary 68
Eleusine coracana 66, 68
elixir 68
end-stage renal disease or ESRD 213
Erectile dysfunction (ED) 213
Essential oils 73, 74
Eucalyptus globulus 70
Euphorbia hirta 72
Evening primrose oil (EPO) 76

F

Fennel 79
Ferula plants 80
Fever 223
Ficus septica 78, 79
Fluid retention 223
Food and Drug Administration (FDA) 63
Forsythia suspensea 82
Foxglove digitalis 84
fruits 222, 228
Fucus vesiculosus 84
Fumiterry 81

G

Gamma-amino-butyric acid (GABA) 97
gamma-linolenic acid (GLA) 76
Garcinia cambogia 98
Gardenia gummifera 93, 94
Garlic 87, 88, 89, 90, 91, 92
Gastroesophageal reflux disease 98
generally recognized as safe (GRAS) 235
Genetically modified organism 100
Glucosyltransferase 100
Glycyrrhetinic acid 97
Glycyrrhizin 62, 63
Gmelina arborea 95, 96
Guinea pigs 100
Gypsophila 34, 35

H

Helianthus annus 102, 103
herb 222
herpes simplex virus (HSV) 159
Hibiscus rosa-sinesis 104, 105
High blood pressure 223, 224
Hordeum sativum 105, 106
Hormone replacement therapy (HRT) 110
Hydrogen cyanide 109
Hydroxycitric acid 108
Hydroxymethylglutarate 109
Hyoscyamus niger 106, 107

I

Immortalitea 114
Immunoglobulins 116
Indian bael 112, 113
Indian vanilla bean 115
inflammatory bowel disease (IBD) 81
Infusion 118
Inhalation 117
Inhibitory concentration 116
Intradermal 118
Intragastric balloon therapy 117
Intrathecal 119
intravenously (IV) 82, 85
Iodine 114
Ipomea aquatic 113
Irish breakfast tea 115
Iron 115
Italian herb seasoning 115, 116

J

Jakhya 127
Jalapenos 129, 130
Jiaogulan 121, 122, 123
Jimbu 126, 127
Juniper 123, 124, 125, 131
Juniperus communis 123, 131
Justicia adhatoda 8

K

Kale 135, 136, 137
Katuk 133, 134
Kava 138, 139, 140, 142
Key lime 48
Kombucha 143, 144, 145

L

Lavender 149, 150, 151, 152
Lecithin 153, 154, 155
Lemon 156, 157, 158, 159, 160, 161, 162, 164
Lemon balm 158, 159, 160, 161
Lemon Myrtle essential oil 163
Leprosy 28, 29
licorice root 166
liverwort 168, 169
Lobelia 170
Lung disorders 223

M

Mahua 173, 174
Marvel grass 171, 172
Menstrual cramps 223
Michelia champaca 178
Mifoil Yarrow 326
Mimosa pudica 176
Moluccas 183
monounsaturated fatty acids (MUFAs) 199
Murraya koenigii 181, 182
Mustard 179, 180

N

neem tree 187
nettle 189, 190
nicotiana tobacco 187
nicotianine 189
Niyog-niyogan 222
nonsteroidal anti-inflammatory drugs (NSAIDs) 20

O

Ochnaceae 191
Olea europaea 198
Olive oil 199
onion 11, 12
Oregano vulgare 201
Organophosphorus chemicals 25

Oryza sativa 196

P

Papaver somniferum 203
Papaya 43
Parenteral delivery 216
Persea americana 204
Phaseolus calcaratus 206
Picrorhiza 205, 206
Pinellia 37
Piper nigrum 209
polycystic ovary syndrome or PCOS 214
pomegranate 213, 214
premenstrual dysphoric disorder (PMDD) 218
Premenstrual syndrome (PMS) 143, 147
Prostaglandins (PGs) 217
Protein kinases 218, 221
Pyrrolizidine alkaloids (PAs) 214

Q

Quebracho 223, 224
Queen's Delight 226, 227
Quince 228, 229, 230

R

raspberry ketone 234, 235
red clover 237, 238, 239
renin–angiotensin system (RAS) 25
Respiratory syncytial virus (RSV) 248
reverse transcriptase (RT) 250
Rheumatoid arthritis (RA) 213
Rhodiola rosea 239, 240, 241
Ricinus communis 231
rose hip 243, 244
Rosemary 53, 244, 245, 246
Royal jelly 246, 247, 248

S

S-adenosylmethionine 261
Sandalwood 252, 253
Sapindus rarak 255
Saraca indica 255, 256
Scientific Committee on Food (SCF) 64
Seasonal affective disorder (SAD) 260
Senna alata 44, 45
Sesquiterpene lactones (SLs) 263
solid-phase microextraction gas chromatography-mass spectrometry (SPME GC–MS) 169
Spasms 223
Stachytarpheta jamaicensis 257
Sugarcane 254

T

Tamarind 269, 270
TCM (Traditional Chinese Medicine) 275
Terminalia bellirica 272
THC, or tetrahydrocannabinol 277
thiobarbituric acid reactive substance (TBARS) 195
toothache-relieving gargle 222
Tribulis Terrestris 274

U

Unicorn root 282
urinary tract infections (UTI) 85
Usnea 284, 285

Uva ursi 281, 282
Uzara 285, 286

V

Valeriana officinalis 287
Veronica anthelmintica 291, 292
Vigna sinensis 293, 294
Vitex agnus-castus 289, 290
Vitex negundo 295

W

Walnuts 297, 299
Wasabi 299, 300
Watercress 301, 302, 303
Wheatgrass 303, 304
Whey 305, 306
white oak 307, 308
Wild yam 309, 310
Withania somnifera 313, 314
Wolff-Parkinson-White (W.P.W.) 315
World Health Organization (WHO) 64
Wormwood 311, 312

X

X Heucherella 319
Xylia kerii 316, 317
Xylopia aethiopica 318

Y

Yarrow 321, 322, 325, 326
Yellow dock 326, 327
Yellow Jessamine 328, 329
Yerba mansa 329, 330
Yerba mate 323, 324
Ylang ylang 324, 325
Yohimbe 331, 332
Yucca 333, 334

Z

Zea mays 335, 336, 337
Zedoary 343, 344
zero moisture basis 342
Zinc 340, 341
Zingiber officinale 338, 339